SciGRID International Conference on Power Grid Modelling 2017

Journal of Physics: Conference Series Volume 977

Oldenburg, Germany
30 – 31 March 2017

ISBN: 978-1-5108-5954-8
ISSN: 1742-6588

Printed from e-media with permission by:

Curran Associates, Inc.
57 Morehouse Lane
Red Hook, NY 12571

proceedings.com

Some format issues inherent in the e-media version may also appear in this print version.

Printed by Curran Associates, Inc. (2018)

For permission requests, please contact the Institute of Physics at the address below.

Institute of Physics
Dirac House, Temple Back
Bristol BS1 6BE UK

Phone: 44 1 17 929 7481
Fax: 44 1 17 920 0979

techtracking@iop.org

Additional copies of this publication are available from:

Curran Associates, Inc.
57 Morehouse Lane
Red Hook, NY 12571 USA
Phone: 845-758-0400
Fax: 845-758-2633
Email: curran@proceedings.com
Web: www.proceedings.com

Table of contents

Volume 977

SciGRID International Conference on Power Grid Modelling

30–31 March 2017, Oldenburg, Germany

Accepted papers received: 20 February 2018
Published online: 12 March 2018

Preface

Papers

IOP Conf. Series: Journal of Physics: Conf. Series **977** (2018) 011001 doi:10.1088/1742-6596/977/1/011001

SciGRID International Conference on Power Grid Modelling

Preface

The **SciGRID International Conference on Power Grid Modelling** was held in Oldenburg (Germany) on the 30th and the 31st March 2017. The conference was organized in the context of the **SciGRID** project financed by the German Federal Ministry of Education and Research (BMBF) within the funding initiative "Zukunftsfähige Stromnetze". The **SciGRID** project was initiated, managed and performed by the **DLR Institute of Networked Energy Systems** in Oldenburg (formally NEXT ENERGY – EWE Research Centre for Energy Technology).

The focus of the **SciGRID** project is to develop methods for the automated generation of power transmission data using open source models and datasets. The main source of input data used in the project is the ***OpenStreetMap*** (www.openstreetmap.org) databank. The models developed in the **SciGRID** project as well as their source code and datasets were made freely available. For more details and information about the **SciGRID** project please refer to the website: www.scigrid.de. For the purpose of the **SciGRID International Conference on Power Grid Modelling**, datasets for transmission networks for the German and the European grids were made available online.

The **SciGRID International Conference on Power Grid Modelling** aimed at presenting the state of the art of the research on power grids and energy system modelling. During two days, experts from the power network and the energy system modelling communities presented and discussed their work. The goal was to share knowledge and help identify common problems encountered in research, industry, and software development topics.

The conference covered a wide range of topics which included:

- Characterization and analysis of power networks
- Comparison & validation of grid models and grid data
- Models of power grid for all voltage levels
- Grid models in energy systems
- Energy market models and simulations
- Transmission and distribution expansion planning
- Open source grid models and grid data
- (optimal) power flow models on (different) grids
- Power grid operations
- Algorithms for power network analysis

The participants to the **SciGRID International Conference on Power Grid Modelling** could either submit a paper and or a poster with their topics and an oral presentation. In total 15 expert talks and 6 posters dealing with many topics related to grid modelling were presented.

These proceedings contain 9 peer-reviewed publications which were presented at the **SciGRID International Conference on Power Grid Modelling**. The goal in publishing the proceedings of the conference as an open source volume is to help disseminate the conference material to a wider audience. Moreover, the proceedings should contribute to making research in the field of power grids modelling more transparent and tractable.

We thank all the conference participants and the authors who contributed to the proceedings. We also warmly thank those who generously accepted to act as reviewers for the contributions presented here.

We are thankful to the **BMBF** Research Initiative **"Zukunftsfähige Stromnetze"** for the funding of the **SciGRID** project (funding code **03SF0471**) and for the generous funding of the **SciGRID** conference.

Oldenburg, Germany
December 2017
Wided Medjroubi

The list of the talks presented is as follows:

U.P. Mueller, *"The eGo grid model: an open source approach towards a model of the German high and extra-high voltage power grids"*, Flensburg University of Applied Sciences.

D. Mende, *"Combined power market and grid modelling: implementation and applications"*, Fraunhofer Institute for Wind Energy and Energy Systems Technology (IWES).

S. Marwitz, *"Extension algorithm for generic low-voltage networks"*, Fraunhofer Institute for Systems and Innovation Research (ISI).

J. Amme, *"The eGo grid model: an open-source and open-data based multi-voltage synthetic grid model for distribution power supply"*, Reiner Lemoine Institute gGmbH.

J. Hörsch, *"An open source large-scale spatial optimization model of the European electricity system"*, Frankfurt Institute for Advanced Studies (FIAS).

C. Senkpiel, *"Evaluation of load-flow in a linear unit-commitment expansion optimization model"*, Fraunhofer Institute of Solar Energy Systems (ISE).

P. Neetzow, *"The impact of battery storage siting on network congestion: an assessment for Germany"*, University of Oldenburg, Humboldt-Universität zu Berlin.

A. Kies, *"The impact of grid storage application on battery life degradation in electric vehicles"*, Frankfurt Institute for Advanced Studies (FIAS).

C. Matke, *"Resilience analysis in the SciGRID network"*, DLR Institute of Networked Energy Systems.

F. Böing, *"Assessment of grid optimizing measures for the German transmission grid using open source grid data"*, FfE – Forschungsstelle für Energiewirtschaft e.V.

W. Biener, *"Impact of grid reduction on modelling accuracy of line usage-rates"*, Fraunhofer Institute of Solar Energy Systems (ISE).

S. Kettemann, *"Propagation of disturbances in power grids"*, Jacobs University Bremen.

IOP Conf. Series: Journal of Physics: Conf. Series **977** (2018) 011001 doi:10.1088/1742-6596/977/1/011001

S. Auer, *"Topology related instabilities driven by intermittent fluctuations in power grids"*, Postdam Institute for Climate Impact Research (PIK).

P. Lind, *"The role of heterogeneities for the stability of the German power grid"*, Osnabrück University.

The list of the posters presented during the poster sessions is as follows:

K. Schmietendorf, *"The impact of wind feed-in on power system stability and quality"*, University of Oldenburg, ForWind Research Centre for Wind Energy Research.

W. Heitkoetter, *"Introduction of an Open Source Tool for the Characterization and Comparison of Open Source Power Grid Models"*, DLR Institute of Networked Energy Systems, RWTH Aachen University.

M. Conrath, *"Spreading of perturbations in self-similar, tree-like electricity grids"*, Jacobs University Bremen.

K. Karamanlis, *"Linear modeling of the electric power transmission network of Europe using network reduction techniques"*, Institute of Electrochemical Process Engineering (IEK-3), Jülich Research Center GmbH Chair for Fuel Cells, RWTH Aachen University.

P. Schultz, *"Enhancing SciGRID with distibution grids - a network of networks model"*, Potsdam Institute for Climate Impact Research (PIK).

A. Hoffrichter, *"Impact of considering 110 kV grid structures on the congestion management in the German transmission grids"*, RWTH Aachen.

IOP Conf. Series: Journal of Physics: Conf. Series **977** (2018) 011002 doi:10.1088/1742-6596/977/1/011002

Peer review statement

All papers published in this volume of *Journal of Physics: Conference Series* have been peer reviewed through processes administered by the proceedings Editors. Reviews were conducted by expert referees to the professional and scientific standards expected of a proceedings journal published by IOP Publishing.

IOP Conf. Series: Journal of Physics: Conf. Series **977** (2018) 012001 doi:10.1088/1742-6596/977/1/012001

Impact of grid reduction on modelling accuracy of line usage rates

Wolfgang Biener*[1], Charlotte Senkpiel[1], Shivenes Shammugam[1], Klaus René Garcia Rosas[1,2], Manuela Linke[1,3] and Oliver Eibl[3]

[1]Fraunhofer Institute for Solar Energy Systems ISE, Freiburg, Germany
[2]Albert-Ludwigs-Universität Freiburg, Germany
[3]Eberhard Karls Universität Tübingen, Germany

*Wolfgang.Biener@ise.fraunhofer.de

Abstract Grid constraints have a strong influence on the results of energy system optimization models. Since energy system optimization is a problem of high complexity and data e.g. power production potential is not necessarily available for all grid nodes grid reduction is one method to overcome both. It reduces the computational burden and it may fit grid structure to given data.. This paper addresses the influence of grid reduction on line usage rates. Line usage rates are indicators to determine necessary investments for grid reinforcement. Three reduction methodologies, preserving original grid parameters, hence allowing to calculate which line of the original grid needs to be upgraded, are tested. Statistical error measures show that those reduction methodologies lead to high errors for usage rate calculation, due to the influence of the intra-zonal network. The high error rates could be reasoned and approaches were identified which might overcome them.

1. Introduction

Due to the "Energiewende" in Germany an increased use of distributed generation systems will lead to challenging power flow scenarios in the existing grid, which was (initially) optimized for a centrally dominated structure. The need for grid transformation has to be evaluated, analysed and optimized to obtain cost-effective solutions. The ENTIGRIS energy system optimizer therefore is able to quantify such solutions including optimal grid expansion and operation in conjunction with the optimal technology mix, regarding energy sources and storage systems [1]. Due to the complexity of such systems, the simplification of high-dimensioned grid and power flow models is necessary to ensure time-efficient overall optimization processing. Therefore this paper evaluates the effect of spatial grid reduction on power flows. In the base case, a full transmission grid model is calculated. The base case is taken as reference for the comparison of three different grid reduction methodologies. Grid calculation is based on a DC load flow approach. Goals of grid reduction can be summarized as done by [2]:

- the reduced model should reproduce the physical nature of the power system to be replaced as close as possible
- the equivalent should be flexible enough to handle power system status changes and to be used in a wide range of applications

- the equivalent should be compatible with the computational procedures used to solve subsequent problems
- the equivalent should insure that feasible mathematical solutions are obtained
- the reduced network should contain as small a number of nodes as possible

For energy systems it is additionally important to know which lines have to be reinforced due to overloads.
The remainder of this paper is organized as follows. The second chapter starts with a short overview on grid representations in energy system optimizers and grid reduction approaches. Grid reduction approaches are surveyed from the energy system optimizing perspective as well as the electro-technical point of view. Chapter three explains the construction of the use case and chosen grid reduction methodologies. Grid reduction methodologies are the focus of research in this paper. The impacts of grid reduction on grid representation in energy-system optimizers are evaluated by statistical measures. In chapter four results are presented. A critical discussion on the approach and on the limits of the models follows in chapter five.

2. State of the art

This chapter gives an overview on possibilities to represent grids within energy system models and grid reduction methods. Firstly, grid representations in energy system optimizers are discussed. After that, methods on the reduction of electricity grids are shown.

2.1. Grid representation in energy system models

[3] gives a general overview on grid representations in energy system optimizers. First of all, there is the possibility to model the grid as AC power flow model. This is the most accurate possibility to model electrical grids. Therefore requirements regarding grid data and computational complexity are the highest. In order to correctly solve AC power flows, apart from line and transformer data, additional information of flexible AC transmission systems (FACTS) has to be available. To calculate systems states via AC power flows, quadratic problems need to be solved [4]. If the AC power flow problem is integrated into an energy system model, allowing additional discrete grid expansions, this results in a mixed integer non-linear program (MINLP). The conclusion is that for high temporal and spatial resolution data the complexity of the problem is not suitable for real world problems.

Another possibility to model transmission systems with high X/R ratios is the DC load flow model. Due to assuming a constant voltage magnitude and neglecting line resistances, the AC power flow model can be substituted by a linearized model [4]. Combining an energy system optimization model with a DC load flow model transforms the resulting problem again in a MINLP, which is not convex.

The so called transshipment models are the simplest way to take grids in energy system models into account. The only binding relation between real power system operation and the transshipment model is the balance of power at each node of the grid. According to [3], only continuous grid expansion is taken into account in such a model. If the energy system optimizer is linear, the combination of transshipment model and energy system model holds linearity. In general, grid expansion may also be taken as integer variables into account. Then, the overall model is a MILP as shown in [5]. In [5] an entire energy system of a district is modelled using MILP. Grid constraints are taken into account as thermal limitations. This is possible due to the short line lengths, which imply that the first overload in the grid is always thermal. Hence if there is no thermal overload there is no voltage overloads as well. Grid expansion measures are integrated into the model via binary decision variables. Due to the tree-like grid structure the transshipment model finally is a sufficient representation of the real grid. Lastly, the main advantage of this model is the comparatively low computational burden. Additionally, combinations of the upper methods are possible.

The simplest approach to represent a grid is to neglect it. So does the so called copper plate approach used for example in the REMod-D model [6]. This model focusses on the coupling of all sectors. Taking the sector coupling potential into account it aims at answering the question what a 100 % renewable energy supply for German energy infrastructure implies. Within Germany no restrictions for the electrical grid are taken into account. There is only a maximum value for the electricity import defined as a constraint.

2.2. *Grid reduction for grid simulations*

This subchapter gives an overview on grid reduction methodologies that do not aim at integrating grids into energy system optimizers, but at maintaining grid behavior as good as possible for grid simulations.

2.2.1. *Wards method*

One of the first important publications on grid reduction or grid equivalencing are [7] and [8]. Grid analyzers at that time had a constraint on the maximum number of nodes and since power systems were growing faster than grid analyzers' capacity, a method to reduce the complexity of the grid had to be found. From a grid operator's point of view he only needs to know what happens in the zone that he is responsible for. Therefore they developed a method based on AC load flows to reduce the behavior of the external zones on a lower amount of characteristic nodes and lines. For that the interaction with the grid operator's zone does not change significantly.

2.2.2. *REI-DIMO*

Another method to reduce complexity of grid structures is REI-DIMO's method. It is an improvement of REI's method to carry out steady state stability analysis. Similar to Wards method, it is based on AC load flow. In order to reduce the amount of necessary nodes for grid calculation, only manually defined essential nodes are kept within the original grid model. Non-essential nodes are combined to a set of virtual nodes whose behavior regarding power injection is linearized depending on voltage. Updating of grid reductions is necessary depending on system state changes in order to have a high accuracy of system reaction [2].

2.2.3. *Zonal Power Transfer Distribution Factors*

Power Transfer Distribution Factors (PTDFs) represent line usage rate changes for varying node power injections. Zonal-PTDFs are aggregated PTDFs which represent the distribution of power flow when a certain amount of power is transferred from one zone into another. By doing this, parallel flows are taken into account. The accuracy of this method strongly depends on the distance to the working point, therefore PTDFs need to be updated regularly [9]. According to [10] there is even a high influence between day and night operating conditions for certain elements of the PTDF matrix.

2.2.4. *Equivalent impedances method*

In [11] the equivalent impedances method is developed. This method minimizes the squared differences between the flows in the original grid and the reduced grid. The reduced grid is represented as well as DC load flow model. In order to minimize squared flow differences, the admittance matrix of the resulting grid is modified. In order to guarantee a broad operating range, optimization is done for a large set of working points. The resulting grid does not take the possibility of more than one connection between reduced grid nodes into account.

2.3. *The role of clustering for grid reduction methods*

In many cases the grid will be reduced according to clusters. Although the choice of clusters may have great influence on the result after grid reduction, clusters are often predetermined by regions or

IOP Conf. Series: Journal of Physics: Conf. Series **977** (2018) 012001 doi:10.1088/1742-6596/977/1/012001

countries like in [10], [11] and [12]. Adjustments of line parameters or alternative forms of load flow expressions like Zonal-PTDFs are necessary to obtain load flow behaviour. [13] and [14] combine the clustering with the grid reduction procedure to obtain clusters which are suitable for their reduction methods. In our work we will use predefined regions as clusters which are based on areas of transmission system operators. We don't adapt line parameters arbitrarily to reproduce the original load flow like in the equivalent impedances method. Instead we use methodologies which intend to keep important properties of removed lines and integrate them into those lines which are kept after the reduction.

3. Methodology

As seen in the previous section there is a wide range of possibilities to reduce grids. Each of them has their own drawbacks. This is the reason on why we decided to test new possibilities in this paper.
An overview on the methodology of this paper is given in Figure 1. The ENTIGRIS energy system optimization model creates nodal powers with its highest possible spatial resolution. Those nodal powers are transferred to the DC load flow solver. The DC load flow algorithm calculates the power flows within the grid based on this detailed input. Then grid reduction methodologies, introduced in subchapter 3.5. , are applied to reduce the grid to predefined regions. Each region is seen as a virtual node uniting all nodes within that region. All virtual nodes are connected to each other following the specific grid reduction methodology.

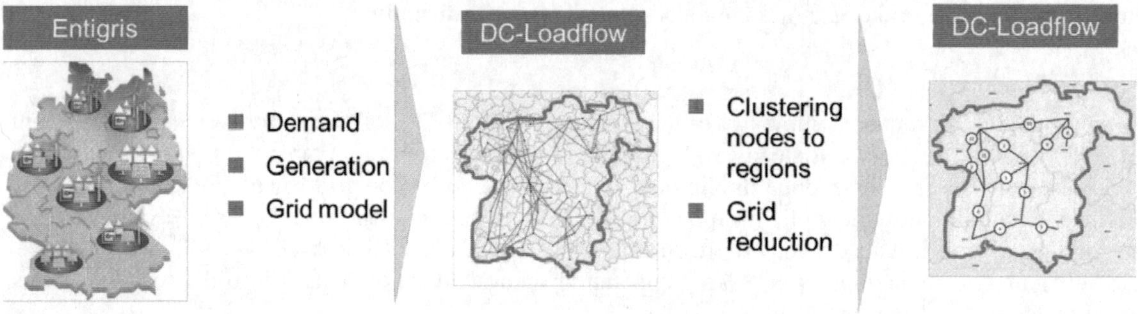

Figure 1. Procedure to test accuracy of grid reduction methodologies.

3.1. Calculation of nodal powers

The ENTIGRIS[1] energy system optimizer calculates power exchanges for each region. Since runtime is long only 120 load cases are calculated. Those load cases include all combinations of high and low PV and wind productions as well as high and low demand and local effects.
The highest possible regional resolution for ENTIGRIS is NUTS3 level due to lack of data for further underlying regions. In a first step the grid needs to be slightly adapted for the ENTIGRIS model to have exactly one grid node in each model region. The following cases make this necessary: NUTS3 regions having more than one node or none at all. Since every region is connected to the transmission grid via underlying grid levels, we combine regions having no nodes to the region with lowest centroid distances. NUTS3 regions with more than one node are defined as one node. This necessitates a post-processing of nodal powers within the DC load flow model. Power of regions having more than one grid node is divided equally to inner nodes. Due to the lack of more precise datasets, a more accurate distribution of power is not possible.

[1] https://www.ise.fraunhofer.de/en/business-areas/energy-system-technology/energy-system-analysis/energy-system-models-at-fraunhofer-ise.html

3.2. Grid data

The ELMOD-DE model is taken as the grid dataset in this paper [12]. According to [12] there is one type of a line trace that may possibly carry two lines. Hence, if there is only one line it may be expanded to two lines. As there are two types of traces there are two different voltage levels 220 kV and 380 kV. 220 kV traces may be upgraded to 380 kV.

3.3. Line parameters, expansion and upgrade costs.

The German grid consists of 436 nodes and 697 lines. For the sake of simplicity we constrain the area to a region including and surrounding Baden-Württemberg. Resulting number of nodes and lines is 69 and 127 respectively. Since there are no security calculations performed, maximum line capacity is reduced to 70 % of its original capacity.

3.4. Load flow calculation and grid reinforcement

Load flows are calculated via a DC load flow model that is integrated into an optimization problem as done in [12]. Due to the formulation as an optimization problem, necessary calculations of line reinforcements due to overloads are integrated within the solution of the load flow. The optimizer finds an angle φ_i for each node i in the grid which ensures balance of power P_i at each node. The angles depend on the lines' susceptance matrix B_{lines} that is constructed from lines between nodes.

$$P_i = \sum_{k=1}^{N} B_{ik} \left(\varphi_i - \varphi_k\right)$$

Power flows over a line linearly depend on the difference of the angle from one side to the other. Due to the linear dependence it can be easily integrated into a linear optimization model.
Overloaded lines get reinforced by the minimum costs so that capacity constrains are not violated provided that angle differences φ remain the same. In order to hold linearity, changes in line properties due to line upgrades may not be integrated in the optimization problem. Nevertheless the load flow will be recalculated after reinforcement measures, so that the resulting optimization is a sequential linear program.

3.5. Grid reduction methodologies

Overall goal of this paper is to find a grid reduction methodology with the least loss of accuracy. The most important goal is to calculate line reinforcement costs as exact as possible. Grid reduction methods described in section 2 are not suitable since they either necessitate AC power flows as basis for grid reduction, line reinforcements may not be referred to existing lines or have other drawbacks.

3.5.1. Cross-border

The first method reduces every region to one node. Only lines connecting areas are taken into account.

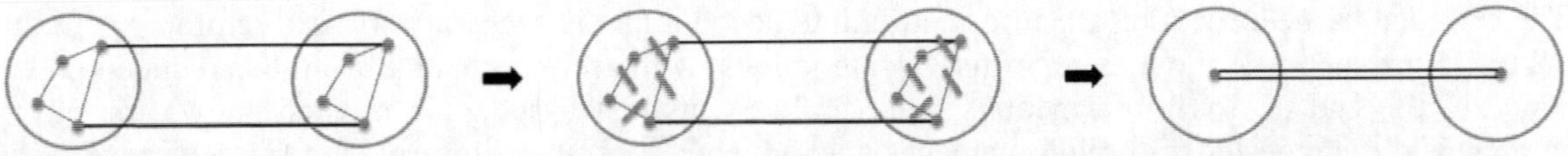

Figure 2. Principle of cross-border methodology. Line lengths of inter-regional lines stay the same.

The advantage of this method is the simplicity of implementation. The disadvantage of this method is the high loss of information from conducting the reduction step. All intra-regional expansions will be lost and expansion costs only take the short lengths of connecting lines into account. This will lead to an underestimation of expansion lengths and costs.

3.5.2. *Equivalent lengths*

The equivalent lengths method overcomes the problem of the cross-border method by distributing the intra-regional line lengths linearly to the cross-border lines according to their lengths.

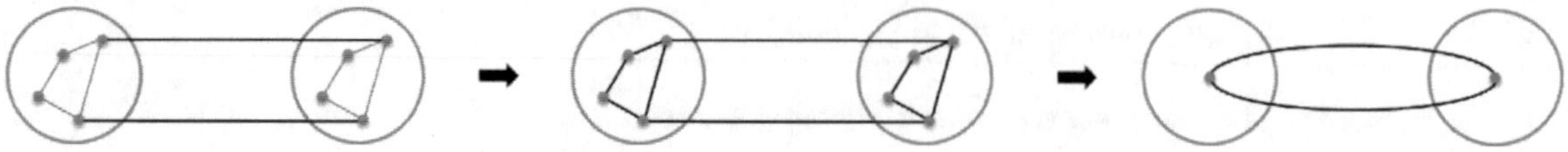

Figure 3. Principle of equivalents lengths methodology. Line lengths of inter-regional lines will be increased artificially.

Consequently if a line needs to be reinforced, costs for expansion rise since longer line lengths are taken into account. In order to reflect the intra-regional behaviour, normalized reactances of intra-regional lines are added to reactances of cross-border lines according to their reactance.

3.5.3. *Shortest paths*

The shortest paths methodology chooses the grid node from the original grid which has the shortest distance to the geographical centre of the combined region.

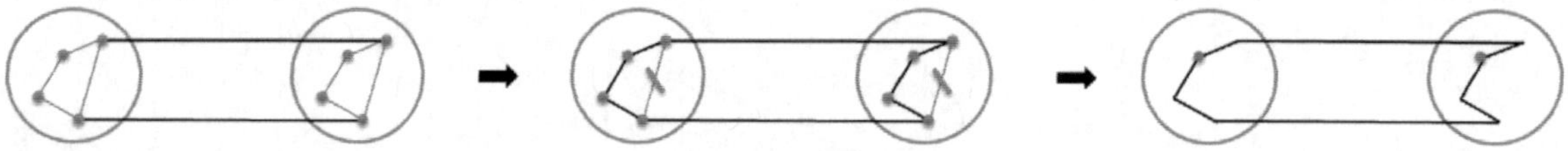

Figure 4. Principle of shortest paths methodology. Line lengths of inter-regional lines will be increased artificially.

Starting from this centre node, the shortest paths to the centre nodes of neighbouring regions are calculated. As distance measure for the determination of shortest paths lines normalized reactance is chosen. Expansion costs for the resulting line are determined as if it would consist out of the original lines. This enables to determine upgrade requirements and costs according to the original line pieces and lengths.

4. Results

In Figure 5 a statistical evaluation of the results of discussed reduction technologies is done. Four errors are taken to analyze differences in line usage. The comparison is done between inter-regional lines of the reference scenario and the lines of the reduced grids. Data for the boxplot is always from a comparison of the lines over all 120 load cases. The first error is the maximum absolute error (MAE). The MAE describes the maximum difference between reference case and reduced grid for a time step. Hence it is a measure for the maximum deviation within a single time step. This basically shows the maximum error which is to be expected for a special load case. Cross-border and equivalent lengths methodologies perform comparable with a high maximum error of about 90 %. Shortest paths method is even worse with an MAE of 250 %.

The maximal usage error takes the information into account that is necessary for grid reinforcement. It subtracts for each line the maximum usage rate values of the reference case from the reduced case. Since in the end grid reinforcement is done for the maximal usage, this method shows the most important difference for grid reinforcement decisions, but it gives no information on quality of grid representation. It is interesting to note that some line usage rates are strongly underestimated for all cases. More importantly it shows that bad results seen from MAE are levelled out by this method for cross-border and equivalent lengths methodologies. Shortest paths remain the worst case here as well.

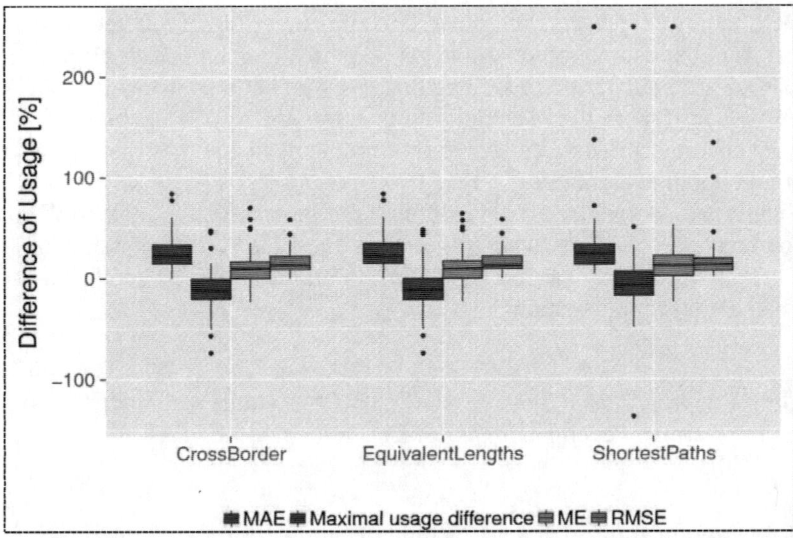

Figure 5. Statistical evaluation of results of all three proposed grid reduction methods. Errors are calculated over each line over 120 defined load cases. MAE: Maximal Absolute Error, ME: Maximal Error, RMSE: Root Means Square Error.

The maximum error ME shows the maximal difference that happens within one load case for a time series. It extends the information given by the MAE on the sign of the maximum error. From the MAE it is not possible to derive if the error is positive or negative. High negative errors basically show which lines are not detected to be overloaded whereas positive errors give a hint if lines are unnecessarily reinforced. In the end it shows that the error within time single load cases is worse for reinforcement evaluation than the maximal usage error but it is better than the MAE for all methodologies but the shortest paths. The reason for the high positive error in the shortest paths method is the loss of many inter-regional lines. The remaining lines basically have to carry the same load as before.

In the end the root-means-square-error (RMSE) shows the overall quality of the usage calculation of reduced cases compared to the reference case. This helps to evaluate, whether the other statistical measures state incidentally good results for reduction methodologies or if the grid is represented well by the reduction in all load cases. The median of all three reinforcement methodologies is at 20 % usage error, the outliers go up to 70 % error and for shortest paths methodology up to 150 %.

5. Discussion and Outlook

The statistical evaluation of the comparison between the reduced grid and the reference grid shows that usage rates calculated on the reduced grid deviate strongly from the reference grid. If the maxima of usage rates are compared, the error is reduced. Since maximal usage rates determine the grid reinforcements, this is an important error measure. A problem of all tested methods is that line characteristics of inter-regional lines are not changed. This leads to an unequal distribution of power flows over lines and to high errors of usage rates. This effect was weakened in the original grid by the intra-regional network. The drawback of unequal power flow distribution is unlikely bigger than the advantage of preserving line characteristics and related reinforcement costs and measures. Hence in future work reactances of parallel inter-regional lines will be equalized. Another reason for the high error rates is that regions to which the original grid is reduced are predefined. Therefore in future work an optimization algorithm will be searched and a sensitivity analysis regarding the definition and number of regions will be carried out.

The replacement of one region with one sole node has been identified as further potential problem, since in some regions nodes are not connected via lines to each other within the region. Combining those nodes to one resulting node can strongly change electrical behaviour of load flows in this grid area. Replacing each unconnected node component inside a region by one node which are not connected to the others in this region can result in a valuable improvement which will be tested in future work.

Although the shortest paths methodology represents the inner grid somehow, representation seems to be that bad, that it leads to a strong over estimation of usage rates. Wards method and the REI-Dimo approach that were described above are pretty good in solving this problem but they necessitate an AC power flow. It needs to be investigated if there is a possibility to integrate them with DC power flows. According to literature Zonal-PTDFs are a method to solve this problem; hence they will be tested against future developed grid reduction methodologies.
The research question of this paper was answered by developing three grid reduction methodologies. For those methodologies error measures were defined to evaluate the effect of grid reduction on energy system optimizations. Their drawbacks can be overcome in future work by modifying the reduction algorithms as described earlier in this section.

6. Acknowledgement

This paper is based upon work in the project AutGrid (No. 03EK3557A), within the initiative future-proof power grids supported by the German Federal Ministry of Education and Research.

References

[1] Kost C P 2014 Renewable energy in North Africa: Modelling of future electricity scenarios and the impact on manufacturing and employment (Ph.D. dissertation Dresden)

[2] Savulescu S C 2002 Solving open access transmission and security analysis problems with the short-circuit currents method Proc. of Latin America Power Conf.

[3] Krishnan V, Ho J, Hobbs B F, Liu A L, McCalley J D, Shahidehpour M, et al. 2016 Co-optimization of electricity transmission and generation resources for planning and policy analysis: review of concepts and modeling approaches Energy Systems vol 7 (Springer) pp 297-332

[4] Glover J D, Sarma M S and Overbye T 2012 Power System Analysis & Design vol 5

[5] Hussein N S and Hübner F 2017 Model based analysis of building refurbishing effects on a decentralized energy system IEWT - Wien

[6] Henning H-M and Palzer A 2014 A comprehensive model for the german electricity and heat sector in a future energy system with a dominant contribution from renewable energy technologies - Part I: Methodology Renewable and Sustainable Energy Reviews 30 1003-18

[7] Deckmann S, Pizzolante A, Monticelli A, Stott B and Alsac O 1980 Studies on power system load flow equivalencing IEEE Transactions on Power Apparatus and Systems PAS-99 2301-10

[8] Ward J B 1949 Equivalent circuits for power-flow studies Transactions of the American Institute of Electrical Engineers vol 68 (American Institute of Electrical Engineers)

[9] Kurzidem M 2010 Analysis of flow-based market coupling in oligopolistic power markets (Ph.D. dissertation Zürich)

[10] Duthaler C, Emery M, Andersson G and Kurzidem M 2008 Analysis of the use of power transfer distribution factors (PTDF) in the UCTE transmission grid Power System Computation Conf.

[11] Anderski T, Surmann Y, Stemmer S, Amprion N, Grisey N, Momot E, Leger A-C, Betraoui B, et al. 2015 E-highway 2050 - Modular development plan of the pan-european transmission system 2050 - European cluster model of the pan-european transmission grid

[12] Egerer J, Gerbaulet C, Ihlenburg R, Kunz F, Reinhard B, von Hirschhausen C, et al. 2014 Electricity sector data for policy-relevant modeling: Data documentation and applications to the german and european electricity markets (Tech. rep. Data Documentation DIW)

[13] Svendsen H G 2015 Grid model reduction for large scale renewable energy integration analyses Energy Procedia pp 349-56

[14] Hörsch J and Brown T 2017 The role of spatial scale in joint optimisations of generation and transmission for European highly renewable scenarios Proceedings of 14th international conference on the european energy market (EEM)

IOP Conf. Series: Journal of Physics: Conf. Series **977** (2018) 012002 doi:10.1088/1742-6596/977/1/012002

Assessment of grid optimisation measures for the German transmission grid using open source grid data

F Böing[1], A Murmann[1], C Pellinger[1], A Bruckmeier[1], T Kern[2], T Mongin[3]

[1] Research Center for Energy Economics (FfE e.V.), Am Blütenanger 71, 80995 Munich
[2] Research Center for Energy Economics (FfE GmbH), Am Blütenanger 71, 80995 Munich
[3] Technical University Munich (TUM), Arcisstraße 21, 80333 Munich & École Supérieure d'Électricité, Grande Voie des Vignes, 92290 Châtenay-Malabry

Abstract. The expansion of capacities in the German transmission grid is a necessity for further integration of renewable energy sources into the electricity sector. In this paper, the grid optimisation measures 'Overhead Line Monitoring', 'Power-to-Heat' and 'Demand Response in the Industry' are evaluated and compared against conventional grid expansion for the year 2030. Initially, the methodical approach of the simulation model is presented and detailed descriptions of the grid model and the used grid data, which partly originates from open-source platforms, are provided. Further, this paper explains how 'Curtailment' and 'Redispatch' can be reduced by implementing grid optimisation measures and how the depreciation of economic costs can be determined considering construction costs. The developed simulations show that the conventional grid expansion is more efficient and implies more grid relieving effects than the evaluated grid optimisation measures.

1. Introduction

Due to an increasing share of renewable energy sources (RES) in the German electricity sector, a secure and reliable energy transmission becomes necessary. The conflict between cost and supply reliability, grid operation and planning faces major challenges. Grid expansion is reliable but unpopular with the public [1] and cost-intensive [2], while several 'Grid Optimisation Measures' (GOMs) are also able to reduce or redistribute transmission load in order to avoid curtailment, redispatch and the construction of new lines. The research project *MONA 2030*[1] analyses and compares the GOMs 'Demand Response' (DR), 'Power-to-Heat' (PtH), 'Overhead Line Monitoring' (OLM) and 'Grid Expansion' (GE) [3]. The FfE energy system model ISAaR[2] is applied for assessing the potential, availability, reliability and cost efficiency of these grid optimisation measures. An evaluation of the measures is carried out for the year 2030 regarding the German transmission grid in consideration of the neighbouring countries. Several approaches are applied to handle OpenStreetMap (OSM) grid data. SciGRID [4], osmTGmod [5] and Gridkit [6] are used as grid data sources beside

[1] MONA 2030 (funding code 03ET4015) is co-funded by German Federal Ministry of Economic Affairs and Energy through the funding initiative "Zukunftsfähige Stromnetze".
[2] ISAaR: Integrated simulation model for planning the operation and expansion of plants with regionalisation.

IOP Conf. Series: Journal of Physics: Conf. Series **977** (2018) 012002 doi:10.1088/1742-6596/977/1/012002

data without an open source license like the BNetzA "Integral" dataset[3] [7] or the grid data sets of the TSOs TransnetBW [8], Amprion [9], 50Hertz [10], Tennet [11] and APG [12][4].
Focus of this paper is the documentation of the modelling methodology, the used grid data, the GOM scenarios and the discussion of simulation results.

2. Method

A valid assessment of GOMs has to be based on quantifiable parameters. Therefore, an approach is developed to meet the requirements of an adequate power flow simulation and a realistic modelling of the GOMs' dispatch. Two major criteria for the planning of grid development are the parameters redispatch and curtailment. Both values call for a comparison of market simulation results and grid based calculations. In order to analyse the impact of grid-orientated GOMs like 'Overhead Line Monitoring' and 'Grid Expansion', a market simulation with low detailed spatial resolution is performed. Thereupon, a so called PTDF (Power Transfer Distribution Factors) run with fixed load and generation data from the market simulation for each node is conducted (see [13] and [14] for reference). While this method is quite suitable for grid-orientated measures, the assessment of measures like 'Power-to-Heat' and 'Demand Response' is more challenging. Their dispatch is influenced by grid restrictions and factors like district heating demand, electricity cost or operational restrictions (e.g. for DR). Therefore, the FfE energy system model ISAaR is extended to perform an integrated dispatch of power plants, renewable energy, Power-to-Heat elements, heat plants and industrial consumers with DR, considering a PTDF-linearised grid consisting of DC and AC lines. This approach allows for performing a uniform comparison among all GOMs regarding the parameters curtailment, redispatch, and economic costs. The coupling of electricity and heat generation within the model benefits the evaluation of combined system costs and facilitates comparisons of GOMs considering effects on the complete, coupled energy system.
The following description of the methodology is divided into three subsections:

1. Description of the ISAaR model with focus on the modelled elements of the energy system.
2. Implementation of a linearised power flow approach, known as PTDF.
3. Discussion of the setup of the optimisation sequence.

2.1. ISAaR Model

The FfE energy system model ISAaR has been developed within the project "MOS – Merit-Order of Energy Storage" [14]. The linear model optimises the deployment and the expansion of power plants. In order to evaluate the previously mentioned GOMs, the models' structure is modified as shown in **Figure 1.**

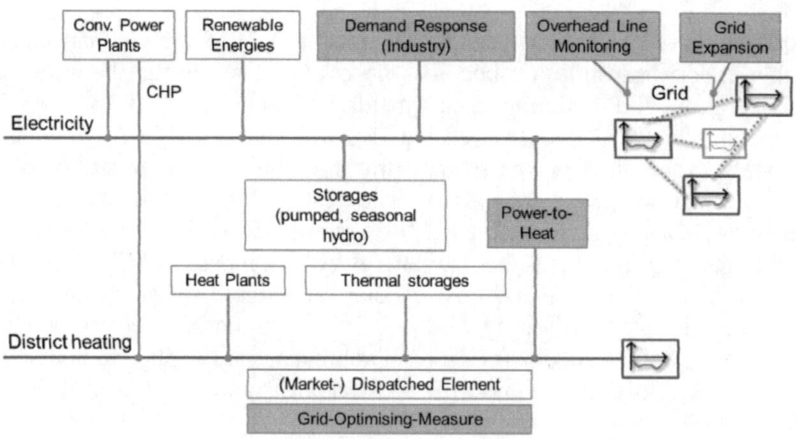

Figure 1. Schematic Overview of the ISAaR models' structure.

[3] This dataset was made available by BNetzA for exclusive project use.
[4] Each dataset is publicly available via download; no license information is given.

The spatial resolution for the electricity sector in the German-Austrian market region is grid nodes (#496) and for the heating sector grid regions (#26). The grid regions are described in [16] for Germany and in [17] for Austria. At each node or in each region a load equation is set for the electrical and the thermal sector. Additionally, the thermal sector is divided into two parts: public demand and industrial demand. This adjustment is caused by major differences in the seasonal characteristics of the load curves.
In general, the model follows the basics of linear programming for energy systems as described in [18] and [19]. Differing from the traditional formulation for power plants and storages, the mathematical formulation for non-typical elements like Demand Response, Power-to-Heat or combined heat and power plants (CHP) are described in detail in [15].

2.2. Technical Implementation
Many energy system modellers use predefined software like GAMS [20], whereas our approach is based on a combination of Matlab®, PostgreSQL, and CPLEX®. Especially the usage of PostgreSQL ensures a high degree of flexibility in generating and combining scenarios. The internally developed "process model" ([15] and [21]) is an important element when working with large amounts of data covering grids, renewable energies, loads and power plants. The technical implementation of the ISAaR model is shown in **Figure 2**.

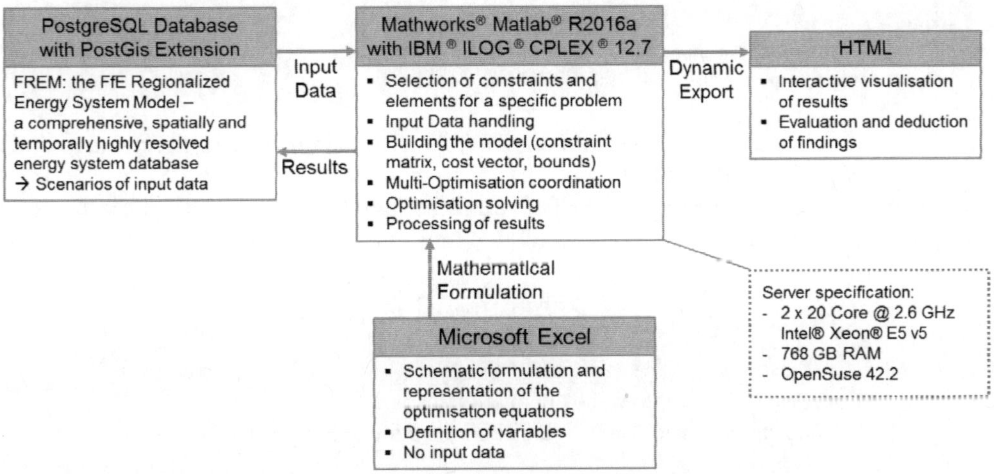

Figure 2. Technical implementation of the ISAaR model.

Standard simulation runs use a rolling horizon (168 h) approach for a whole year and take around 20 to 30 h of computation time. Up to six simulations can be solved simultaneously in parallel configuration. The standard model consists of 496 nodes, applied in a PTDF-grid simulation and the elements mentioned in Figure 1.
Simulation results are visualised interactively by an html-webpage for the evaluation of outcomes and the derivation of insights. A *Leaflet* JavaScript map [22] displays the load of transmission lines and the amount of curtailment and redispatch. Power plant deployment, the utilisation of renewable energy and further analysis is depicted in charts by *amCharts* [23].

2.3. Grid Modelling: DC power flow
Typical grid representations as given in [24] provide good accuracy but come up with computational expenditure. In order to reduce complexity while providing reasonable accuracy, the ISAaR model uses a DC power flow, a linearization of non-linear power flow equations. A clear derivation of this method is given in [25] and [14]. Load flow equations are simplified, assuming no voltage drops, small voltage angles along a transmission line and the disregard of reactive power and line losses. The calculation of the utilisation of overhead lines $\boldsymbol{P_{link}}$ is simplified by a multiplication of the PTDF

matrix (Power Transfer Distribution Factors matrix) and the vector $\boldsymbol{P_{node}}$, representing power injections at the grid nodes:

$$\boldsymbol{P_{link}} = PTDF \cdot \boldsymbol{P_{node}} = (B \cdot A) \cdot (A^T \cdot B \cdot A)^{-1} \cdot \boldsymbol{P_{node}} . \tag{1}$$

A [V]:	Incidence matrix, describing the grid topology
B [1/Ω]:	Diagonal matrix of line susceptances $B = diag(1/x)$
$\boldsymbol{P_{node}}$ [W]:	Vector of power injections at the grid nodes
$\boldsymbol{P_{link}}$ [W]:	Vector of power flows of the overhead lines

Due to its singularity, the expression $(A^T \cdot B \cdot A)$ cannot be inverted. Therefore, one random node of the network has to become a reference point and must be removed from **equation 1**. A power injection in $\boldsymbol{P_{node}}$ can then be interpreted as a transaction $\boldsymbol{T}$ of energy from one node of the network to the declared reference point. The values in a column of the PTDF matrix represent the share of the power from a corresponding power transaction in $\boldsymbol{T}$ flowing through the lines of the network. Equation 1 establishes a causal link between the resulting power flow and the power transaction between nodes. Since the European transmission network consists of transmission lines with different voltage levels a transformation of line parameters for normalisation is applied:

$$X^* = a^2 \cdot X \text{ with } a = 1/u . \tag{2}$$

X [Ω]:	Line reactance
X^* [Ω/V²]:	Normalised line reactance
u [V]:	Voltage level of the line

This transformation integrates all AC lines into one PTDF matrix. High voltage direct current (HVDC) lines have the ability to control the flow of current and are therefore, contrary to AC lines, independent from power injections at the grid nodes. A HVDC line is included into the PTDF matrix by creating a new transaction $T_{x\rightarrow y}$ from line start node x to the line end node y. Its power flow is assigned to $P_{link,DC}$:

$$\begin{bmatrix} \boldsymbol{P_{link,AC}} \\ P_{link,DC} \end{bmatrix} = \begin{bmatrix} PTDF & 0 \\ 0 & 1 \end{bmatrix} \cdot \begin{bmatrix} \boldsymbol{T_{AC}} \\ T_{x\rightarrow y} \end{bmatrix} . \tag{3}$$

$\boldsymbol{T_{AC}}$ [W]:	Transaction of power towards the reference point
$T_{x\rightarrow y}$ [W]:	Transaction of power over a HVDC line from start node to end node

2.3.1. *Accuracy of DC power flow*

The linearisation of the power flow is accompanied with a decreased level of accuracy. In the papers of [26] and [13], the authors show a reasonable precision of the DC power flow if line loads are smaller than 70 % and deviations of voltage levels in the grid are reasonably small.

2.3.2. *Correction of line data*

With the collected grid data, mentioned in chapter 4.1, the grid model is generated and the outcome is revised. Due to the variety of data sources, some lines possess faulty and dissimilar line parameters (especially line reactances), resulting in distorted power flows in the ISAaR model. To overcome this issue, these inconsistencies are rectified in a systematic correction process which is described in [27]. Only obviously wrong parameters are corrected utilising further available line data and applying standard values where necessary.

2.4. Optimisation Sequence

Figure 3 reveals the structure of the models' optimisation sequence which allows for an analysis of the complete European energy system as computational effort is reduced. The two European simulations generate cross-border-flows for further simulations of the German-Austrian energy market which depend on available net transfer capacities between market regions and the power plant capacities in Europe. Therefore, a European market simulation (EU-Ma[5]) is performed first. The cross-border congestion management of this simulation run is based on the NTC (Net Transfer Capacity) approach. Cross-border line projects are considered with the same ratio of NTC to thermal capacity as existing cross-border links. The spatial resolution of load data and renewable generation corresponds to the NUTS 3 level [28]. Power plants are modelled per unit based on the 'Platts power plant database' [29][6] in combination with the TYNDP[7] scenario "Vision 2" [30]. The values for load and installed capacity of renewable energy per country are also derived from [30]. The regionalisation is based on a geospatial analysis in combination with weather data (7 km grid) from the "COSMO-EU" dataset [31] and statistical data like Eurostat [32]. Further information and citation of used data can be found in [27].

For the following step (EU-Ne[8]), the aggregated European transmission grid (see section 4.1.) is added to the model. Due to the high number of nodes (#1500) and lines (#2800), resulting in exceeding computation capacities, a simple PTDF run is performed with a fixed plant dispatch. The optimised elements are HVDC lines, synthetic generation plants and synthetic consumers to keep every load equation feasible. The dispatch of those synthetic elements may be interpreted as (cross-border) demand of curtailment and redispatch.

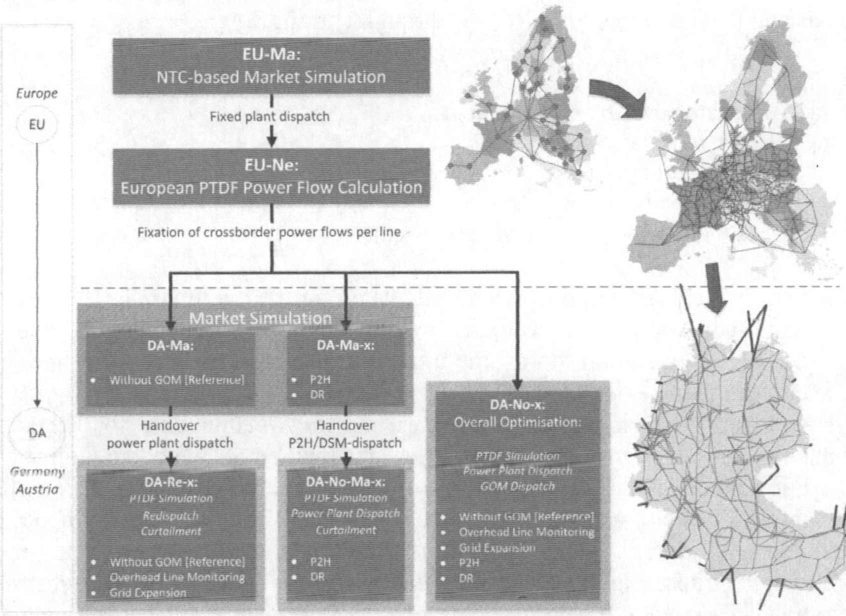

Figure 3. Schematic representation of the optimisation sequence.

With fixed, previously calculated European cross border flows, a market simulation for the German-Austrian market region is performed (DA-Ma). To compare market and grid related dispatch of the GOMs Power-to-Heat and Demand Response, a market simulation considering these GOMs (DA-Ma-x[9]) and a PTDF run with fixed market dispatch of GOMs (DA-No[10]-Ma-x) is performed. For grid related GOMs ('Overhead Line Monitoring' and 'Grid Expansion'), a PTDF run with additional costs on deviation from market dispatch is set up in order to quantify curtailment and redispatch volumes (DA-Re-x). It is not possible to determine the redispatch volume if a load-manipulating GOM like

[5] Ma: Market
[6] The 'Platts power plant database' is a commercially available dataset.
[7] Ten-Year Network Development Plan
[8] Ne: "Netz" (Grid)
[9] „x" is a placeholder for different GOMs.
[10] No: "Netzoptimierung" (grid optimisation)

'Demand-Response' or 'Power-to-Heat' is used. Therefore a third PTDF run is carried out. This run contains an optimised power plant dispatch and GOM dispatch. The resulting overall economic costs for power and heat generation from two simulations with and without the GOM are taken to assess the GOMs impact.

3. Grid Optimising Measures (GOMs)

The impact of a GOM is strongly dependent on its level of implementation. For example, the grid expansion of multiple lines may lead to higher specific savings than expanding one single line. In order to consider this effect, several scenarios are formed to represent different implementation levels of the GOMs 'Overhead Line Monitoring' and 'Grid Expansion'. The selection of lines to be upgraded or monitored first, is based on their line loading in the reference case.

3.1. Overhead Line Monitoring (OLM)

'Overhead Line Monitoring' allows to recognize and exploit additional existing weather-related transmission capacities. The temperature of a line is monitored in order to control its slack and thus keep to valid norms (DIN EN 50341) while avoiding early ageing [33]. With equations for the thermal balance in an overhead line discussed in [3], the line temperature can be approximated dependent on weather conditions and the heat input $(I^2 \cdot R(T_C))$ through current flow:

$$q_c + q_r = q_s + I^2 \cdot R(T_C)\,. \tag{4}$$

q_c [W/m]: Heat removal through convection
q_r [W/m]: Heat removal through radiation
q_s [W/m]: Heat input through solar radiation
I [A]: Current through power line
$R(T_C)$ [Ω/m]: AC resistance of power line at line temperature T_C

The maximum current in the overhead line can be calculated for the maximum line temperature T_C and the prevailing weather-conditions wind velocity, wind direction, solar radiation, and ambient air temperature. Divided by the maximum norm-line transfer capacity of the overhead line, its additional transmission capacity can then be depicted.

Knowing the route of all German transmission lines and the weather data for the year 2012, the available potential of additional transmission capacities is computed for every single line and for every hour of the year. Furthermore, technical restrictions of OLM are considered: on the one hand, current flow is restricted through limitations in elements like circuit-breakers. On the other hand, current flow has to be limited due to a higher voltage drop and the increased need of reactive power in the lines, leading to a risk of system instabilities. Thus, additional transmission capacities are restricted to 50 % of the norm line transfer capacity, except the current flow is restricted further by technical limits of other elements.

3.2. Power-to-Heat (PtH)

As shown in **Figure 1**, Power-to-Heat (PtH) is a sector coupling element. The dispatch of this element depends on the thermal load situation in its assigned district or industrial heating network. On the one hand, PtH is used as an additional load which may avoid curtailment of renewable energy if placed at a convenient location within the transmission grid. On the other hand, PtH, in combination with CHP plants and thermal storages, leads to an increased system's flexibility.

Provided that future PtH expansions are not driven by grid related reasons, the penetration rate is set to 25 % of the thermal power input of existing district heating networks (state 2015). Assuming that the largest German district heating networks are equipped with PtH-devices by 2030, 9.2 GW of PtH capacity is expected to be installed.

3.3. Demand Response (DR)

Power demand of the industry sector can be flexibilized and thus be utilised for temporarily unloading bottlenecks in the transmission network. In [34], the technical potential of demand response is

IOP Conf. Series: Journal of Physics: Conf. Series 977 (2018) 012002 doi:10.1088/1742-6596/977/1/012002

assessed and its economic considerations are elaborated. In the ISAaR model, 2.2 GW in large electricity-consuming companies and 1.6 GW in cross-sectional processes are available for flexible usage. In general, the two fields of DR-application, modelled in ISAaR, are the shift and the loss of production. Conditions on the frequency of demand response dispatch are set in [15]. The regionalisation of DR is oriented to the allocation of employee-numbers and heavy industry locations.

3.4. Grid Expansion

Besides 'Overhead Line Monitoring', 'Grid Expansion' (GE) is another approach to add transmission capacities to the grid. 'Grid Expansion' can be realised in several ways. There are the options of constructing new transmission lines in used or new paths, adding circuits to an existing line or increasing its voltage level. Depending on the individual project, grid planners have to choose the best expansion method for an appropriate grid operation. GE usually comes with high investment costs, but also with high benefits for the transmission grid due to decreased curtailment of renewable energy and reduced redispatch volume.
In the reference scenario, an analysis of all transmission lines is conducted. The most stressed lines are then considered for 'Grid Expansion'- measures in several scenarios. For 220 kV AC lines, the voltage level is increased to 380 kV. For 380 kV, new AC lines with a norm voltage of 380 kV are built with two circuits on the existing path.

4. Input Data

Due to the enormous quantity of input data, the focus of this chapter addresses grid data. Further information on input data can be obtained in [35] (scenario "standard").

4.1. Grid Data

Figure 4 gives an overview of the used grid data. In the ISAaR model, the open-source toolkit Gridkit [6] is used to describe the transmission grid of Europe outside Germany and Austria. With an aggregation process, the grid is simplified in countries distant from and not neighbouring Germany and Austria (see [27]). Due to a lack of reasonable line parameters, standard values from literature are used. With results from the SciGRID project [4], grid data in Germany is validated. Applying the data from the model osmTGmod [5], precise geographic information of transmission lines can be obtained and utilised for the computation of additional transmission capacities with OLM. In parts of the grid, where the transmission capacity of the distribution grid becomes relevant (especially the Ruhr area and the region at the German-Austrian border), secondary lines (110 kV) are considered. Line data is directly taken from Open Street Map [36]. Detailed information on the collection of grid data is given in [27].

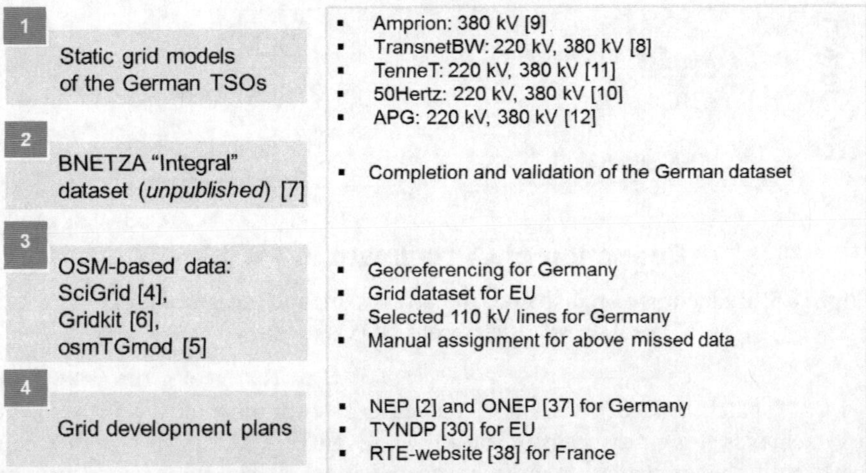

Figure 4. Origin and usage of different grid data sources.

IOP Conf. Series: Journal of Physics: Conf. Series **977** (2018) 012002 doi:10.1088/1742-6596/977/1/012002

4.2. Surrounding Scenario

To perform simulations of the unit commitment in 2030, assumptions on the energy system have to be made. Therefore, the "standard" surrounding scenario of 2030 is set in [35]. The installed capacity of power plants in Germany in 2030 is made up of 23.2 GW coal-fired plants and 34.2 GW natural gas-fired generators. The renewable capacities consist of 73.8 GW onshore wind power, 15 GW offshore wind power, and 75 GW photovoltaic power. The regionalisation of RES is carried out as shown in [39]. The annual energy demand is set to 496.2 TWh_{el}, comprising the conventional load, the demand from heat pumps and the energy demand of the electro-mobility sector. Altogether, the share of RES accounts for 61 % of the final electrical energy consumption.

5. Results

Considering the framework conditions, the gathered cross border flows in Chapter 2.4 and the defined surrounding scenario in Chapter 4.2, simulations of the unit commitment in 2030 are conducted for the market area Germany and Austria. First, a simulation without any GOM is computed and taken as the reference scenario for comparisons with further simulations. Thereafter, several GOMs are implemented for following simulations: two different configuration levels of 'Overhead Line Monitoring' (OLM 1 and OLM Max; 256 km and 23,621 km), a scenario of 'Grid Expansion' (GE; 507 km), a scenario with 'Power-to-Heat' elements and a scenario with 'Demand Response'. The main findings of those simulations are given in the following.

The reduction of redispatch and curtailment are two major criteria when assessing the benefits of a GOM. Compared to the reference scenario, the reduction of redispatch is 0.12 TWh and 0.59 TWh for the two different stages of OLM, the decrease of curtailment amounts to 0.18 TWh and 0.31 TWh (see **Figure 5**).

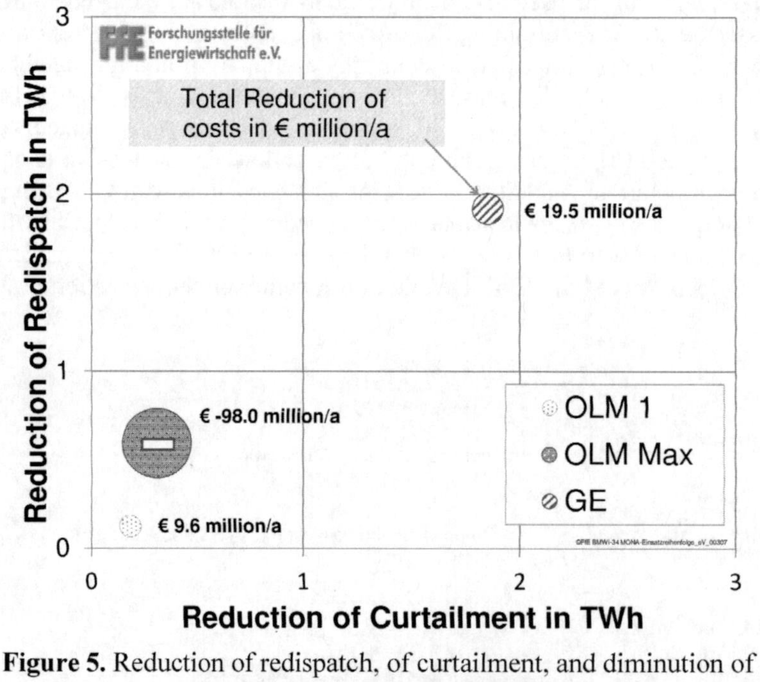

Figure 5. Reduction of redispatch, of curtailment, and diminution of total costs for different GOMs.

With a reduction in redispatch by 1.8 TWh and a decrease of curtailment by 1.9 TWh, the 'Grid Expansion' scenario achieves a greater benefit than OLM. Even the maximum possible equipment with OLM cannot benefit the transmission grid as much as the 'Grid Expansion'-scenario. The decrease of redispatch and curtailment consequently leads to reduced power plant commitment costs due to a more reasonable power plant dispatch and higher integration of RES. Considering the annual costs for the construction of the GOMs, in **Figure 5**, the annual benefits of the measures can be examined. OLM 1 and 'Grid Expansion' come with a reduction in total cost[11] of € 9.6 million and € 19.5 million, respectively. The measure OLM Max causes an additional total cost of € 98.0 million to the system. An intensive expansion using OLM is neither beneficial for the total cost nor for a significant reduction of redispatch and curtailment.

Figure 6 depicts the reduction of curtailment of the GOMs 'Power-to-Heat' with 1.5 TWh and 'Demand Response' with 0.1 TWh. DR has a limited potential to integrate surplus energy from RES due to a restricted shift potential of industrial production processes. However, PtH can permanently integrate excess energy to the heat sector. The reduction in power plant commitment costs due to the GOMs, shown in **Figure 7**, is at € 69 million for PtH and € 49 million for DR. Considering the costs for construction and the total reduction of costs, DR (€ 46 million savings) is more profitable than PtH (€ 11 million). Here, the transfer of load demand to dates with a high share of RES and therefore low prices leads to lower unit commitment costs. Due to the unfavourable spatial arrangement and the restricted shift potential of DR the impact on the network loading is modest.

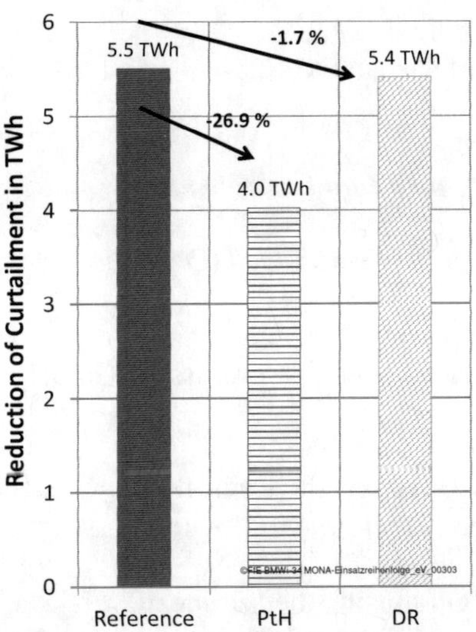

Figure 6. Reduction of curtailment for the measures PtH and DR in comparison to the reference scenario.

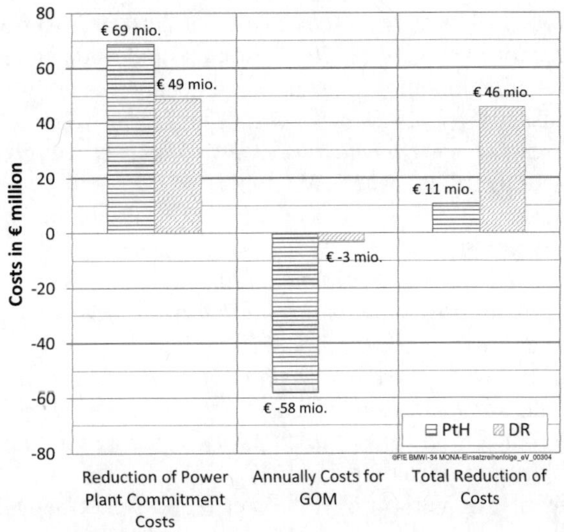

Figure 7. Reduction of unit commitment costs, annuity costs for construction, and reduction of total costs for the measures PtH and DR in comparison to the reference scenario.

6. Conclusion

This paper describes the extension of the ISAaR model with a linearised grid model, enabling a holistic assessment of grid optimisation measures and providing comparability among them. Grid data is gathered from sources of grid operators and open-source data, based on Open Street Map. Different OSM-based toolkits help to achieve a reasonable grid representation.

Within the research project *MONA 2030* (Merit Order Grid Expansion 2030), the ISAaR model is applied for evaluating the grid optimisation measures 'Overhead line monitoring', 'Power-to-Heat', 'Demand Response', and conventional 'Grid Expansion'. In Chapter 5, first results and findings are shown, while further outcomes are supposed to follow from the mentioned project.

[11] Total costs are the costs of one year for the dispatch of power and heat plants minus the annually costs for construction of Grid Optimising Measures.

References

[1] Krack J, Köppl S and Samweber F 2017 Die Akzeptanz des Netzausbaus in Deutschland *et - Energiewirtschaftliche Tagesfragen* Heft **1**/2 (Essen: etv Energieverlag GmbH)

[2] German transmission grid operators 2015 *Netzentwicklungsplan Strom 2025 Version 2015 - Zweiter Entwurf*

[3] Samweber F, Köppl S and Bogensperger A 2016 *Projekt Merit Order Netz-Ausbau 2030 - Teilbericht Maßnahmenklassifizierung* (München: Forschungsstelle für Energiewirtschaft e.V)

[4] Medjroubi W and Matke C 2015 *SciGRID Open Source Transmission Network Model - USER GUIDE V 0.2* (Oldenburg: SciGRID)

[5] Scharf M and Nebel A 2016 *osmTGmod 0.1.1 - Documentation, Version: 0.1.0* (Wuppertal: Wuppertal Institut)

[6] Wiegmans B 2016 *Improving the Topology of an Electric Network Model Based On Open Data* (Groningen, NL: University of Groningen)

[7] BNetzA 2014 *Daten nach §12f Abs.1 EnWG 2013 - Studiennetzmodell des Netzentwicklungsplans 2013* (Bonn: Bundesnetzagentur für Elektrizität, Gas, Telekommunikation, Post und Eisenbahnen)

[8] TransnetBW 2015 *Data for the static grid model https://www.transnetbw.de/de/strommarkt/engpassmanagement/standards-zukunft* (Stuttgart: TransnetBW GmbH)

[9] Amprion 2015 *Static grid model http://www.amprion.net/statisches-netzmodell* (Dortmund: Amprion GmbH)

[10] 50Hertz 2015 *Static grid model http://www.50hertz.com/de/Anschluss-Zugang/Engpassmanagement/Statisches-Netzmodell* (Berlin: 50Hertz Transmission GmbH)

[11] TenneT 2015 *Static grid model http://www.tennettso.de/site/Transparenz/veroeffentlichungen/statisches-netzmodell/statisches-netzmodell (Abrufdatum: 19.03.2015)* (Bayreuth: TenneT TSO GmbH)

[12] APG 2015 *Static grid data https://www.apg.at/de/netz/anlagen/leitungsnetz* (Wien: APG Austrian Power Grid (APG))

[13] Brown T, Schierhorn P P, Tröster E and Ackermann T 2016 Optimising the European transmission system for 77% renewable electricity by 2030 *IET Renewable Power Generation* V **10** I 1

[14] Hagspiel S, et al. 2014 Cost-optimal power system extension under flow-based market coupling *Science direct energy* **66** p 654-66

[15] Pellinger C, Schmid T, et al. 2016 *Merit Order der Energiespeicherung im Jahr 2030 - Hauptbericht* (München: Forschungsstelle für Energiewirtschaft e.V.)

[16] German transmission grid operators 2008 *Übersicht über die voraussichtliche Entwicklung der installierten Kraftwerksleistung und der Leistungsflüsse in den Netzgebieten der deutschen Übertragungsnetzbetreiber - Regionenmodell „Stromtransport 2012"*

[17] Boxleitner M, et al. 2011 *Super-4-Micro-Grid - Nachhaltige Energieversorgung im Klimawandel* (Wien: TU Wien, ESEA/EA)

[18] Weber C 2005 *Uncertainty in the Electric Power Industry - Methods and Models for Decision Support* (Boston: Springer Science + Business Media, Inc.)

[19] Babrowski S 2014 *Bedarf und Verteilung elektrischer Tagesspeicher im zukünftigen deutschen Energiesystem – Dissertation* (Karlsruhe:Fakultät der Wirtschaftswissenschaften am Karlsruher Institut für Technologie)

[20] https://www.gams.com/

[21] Köppl S, Böing F and Pellinger C 2015 Modeling of the transmission grid using geo allocation and generalized processes *ISESO 2015 Heidelberg*

[22] http://leafletjs.com/

[23] https://www.amcharts.com/

[24] Andersson G 2011 *Power System Analysis - Power Flow Analysis Fault Analysis Power System Dynamics and Stability* (Zürich: Eidgenössisch Technische Hochschule Zürich)
[25] Van Den Bergh K, Delarue E and D'haeseleer W 2014 *DC power flow in unit commitment models in: TME Working Paper - Energy and Environment* (Leuven: KU Leuven Energy Institute)
[26] Purchala K, Meeus L, Van Dommelen D and Belmans R 2006 *Usefulness of DC Power Flow for Active Power Flow Analysis* (Leuven: KU Leuven)
[27] Köppl S, Samweber F, Bruckmeier A, Böing F, Hinterstocker M, Kleinertz B, Konetschny C, Müller M, Schmid T and Zeiselmair A 2017 *Projekt MONA 2030: Grundlage für die Bewertung von Netzoptimierenden Maßnahmen - Teilbericht Basisdaten* (München: Forschungsstelle für Energiewirtschaft e.V.)
[28] Eurostat *2007 Regionen in der europäischen Union – Systematik der Gebietseinheiten für die Statistik* – NUTS 2006/EU-27. (Luxemburg: Eurostat)
[29] Platts 2014 *WEPP Database (Europe)* (Washington, DC 20005 USA)
[30] 2015 *Ten-Year Network Development Plan 2016 (TYNDP)* (Brüssel: ENSTO-E)
[31] Schulz J P and Schättler U 2014 *Kurze Beschreibung des Lokal-Modells Europa COSMO-EU (LME) und seiner Datenbanken auf dem Datenserver des DWD* (Offenbach: Deutscher Wetterdienst)
[32] Eurostat 2015 *Energy - Main Tables http://ec.europa.eu/eurostat/web/energy/data/main-tables* (Luxemburg: Eurostat)
[33] Jaufer S, Muhr H M and Schwarz R 2006 *Alterung von Freileitungen Symposium Energieinnovation* (Graz: Technische Universität Graz)
[34] Pellinger C, Schmid, T, et al. *Merit Order der Energiespeicherung im Jahr 2030 - Teilbericht: Technoökonomische Analyse Funktionaler Energiespeicher* (München: Forschungsstelle für Energiewirtschaft e.V.)
[35] Regett A, Zeiselmair A, Wachinger K and Heller C 2017 *Projekt Merit Order Netz-Ausbau 2030 - Teilbericht Szenario-Analyse* (München: Forschungsstelle für Energiewirtschaft e.V)
[36] https://www.openstreetmap.org
[37] Feix O and Hörchens U 2015 *Offshore-Netzentwicklungsplan 2025, Version 2015 - Erster Entwurf der Übertragungsnetzbetreiber* (Berlin: CB.e Clausecker Bingel AG)
[38] 2015 *Shéma décennal de développement du réseau - Édition 2015* (La Défense: Réseau de Transport d'Electricité (RTE))
[39] Schmid T, Gallet M, Carr L and Jetter, F 2015 *Regionalisierung der dezentralen Stromerzeugung im Netzentwicklungsplan 2016 - Methodik und Ergebnisse* (München: Forschungsstelle für Energiewirtschaft e.V.)

IOP Conf. Series: Journal of Physics: Conf. Series **977** (2018) 012003 doi:10.1088/1742-6596/977/1/012003

The eGo grid model: An open source approach towards a model of German high and extra-high voltage power grids

Ulf Philipp Mueller[a], Lukas Wienholt[b], David Kleinhans[c], Ilka Cussmann[a], Wolf-Dieter Bunke[d], Guido Pleßmann[e] and Jochen Wendiggensen[a]

[a] Flensburg University of Applied Sciences, Centre for Sustainable Energy Systems, Kanzleistraße 91-93, 24943 Flensburg, Germany
[b] NEXT ENERGY EWE Research Centre for Energy Technology at the University of Oldenburg, Carl-von-Ossietzky-Str. 15, 26129 Oldenburg, Germany
[c] Scheideweg 9a, 26121 Oldenburg
[d] Europa-Universität Flensburg, Centre for Sustainable Energy Systems, Auf dem Campus 1, 24943 Flensburg, Germany
[e] Reiner Lemoine Institut gGmbH, Rudower Chaussee 12, 12489 Berlin, Germany

E-mail: `ulf.p.mueller@hs-flensburg.de`

Abstract.
There are several power grid modelling approaches suitable for simulations in the field of power grid planning. The restrictive policies of grid operators, regulators and research institutes concerning their original data and models lead to an increased interest in open source approaches of grid models based on open data. By including all voltage levels between 60 kV (high voltage) and 380 kV (extra high voltage), we dissolve the common distinction between transmission and distribution grid in energy system models and utilize a single, integrated model instead. An open data set for primarily Germany, which can be used for non-linear, linear and linear-optimal power flow methods, was developed. This data set consists of an electrically parameterised grid topology as well as allocated generation and demand characteristics for present and future scenarios at high spatial and temporal resolution. The usability of the grid model was demonstrated by the performance of exemplary power flow optimizations. Based on a marginal cost driven power plant dispatch, being subject to grid restrictions, congested power lines were identified. Continuous validation of the model is nescessary in order to reliably model storage and grid expansion in progressing research.

1. Introduction

Professional grid planning undertaken by grid operators and research institutions has been highly restricted and proprietary in past and recent times [1]. In contrast, the research project *open_eGo*[1] aims at developing a transparent grid planning tool which exclusively uses open source and open data methods. This contribution provides a detailed description of the grid model developed within the project. The grid model focuses on the German high and extra-high voltage (HV and EHV) level. The renewable energy (RE) capacity is mostly installed in the

[1] `https://openegoproject.wordpress.com/`

distribution grid [2] inducing changes from unidirectional to complex bidirectional power flows. This implies that the 110kV grid is increasingly loaded in contrast to its original dimensioning. Consequently, considering its high degree of interconnection, the grid originally serving solely for distribution, increasingly functions as a transmission grid. Therefore the integrated modelling of EHV and HV level has become very relevant in the context of the transition of the German energy system [3, 4]. This integral approach features a major difference to comparable approaches in the research area (e.g. [5, 6, 2]). Moreover, in order to model effects concering the German energy transition (in 2050 80 % of the demand shall be produced by RE [7]) it is necessary to consider the status quo as well as the expected future development of the energy supply.In addition, the modelling of the medium and low voltage level is an immanent part of the project and is covered in an adjoined contribution (see [8]).

Consequently, the following research questions shall be the guideline for the approaches presented in this work.

- Is it possible to integrate HV and EHV grid data into one plausible grid model by utilizing open data and open source methods?
- How can the different data sources for grid topology, generation and demand be merged into one spatially and temporally consistent data set concerning status quo and future scenarios?
- How does the calculation of power flows perform for the developed model? Is it possible to produce plausible dispatch-optimized power flows?

Future contributions affiliated to the project *open_eGo* will focus on the optimization of eventual investment decisions concerning storage and grid expansion. This work describes the exogenous setting for a further elaboration on these endogenous variables.

The development of a grid topology model based on OpenStreetMap (OSM)[2] is challenging especially when integrating the 110 kV grid. An approach based on the tool osmTGmod[3] is described in the next section. Several assumptions on electrical parameters have to be defined to allow power flow simulations (Section 3). Decoupled from the electric grid, the overall constitution of generation and demand is characterized for a status quo scenario and two future scenarios (Section 4). In Section 5, it is described how generators and loads are spatially allocated and connected to the grid topology. Subsequently, the definition of the temporal resolution is outlined (Section 6). Consequently, the methodologies[4] lead to a grid data model[5] which was exemplarily applied within a Linear Optimal Power Flow (LOPF) method deriving results for power plant dispatch optimization subject to grid restrictions.

2. A grid topology model based on osmTGmod

The generation of a grid topology can be considered as a necessary first step in creating a complete grid model ready for power flow simulations. Due to restrictive data policies, German grid operators only publish very basic information and data on their grid models. While the TSOs publish static grid models of their respective EHV areas, on an HV level the DSOs only publish maps indicating the geographical position of assets, but no information on the technical parameters. Apart from these publications by grid operators, there is the ENTSO-E map compromising the EHV infrastructure in Europe. While all this information could theoretically be made usable for power flow simulations with some effort, the most important factor against

[2] `http://www.openstreetmap.org`
[3] `https://github.com/openego/osmTGmod`
[4] The implementation is published under an open license (*AGPL* `https://www.gnu.org/licenses/agpl-3.0.en.html`) and can be accessed at a public repository (`https://github.com/openego/data_processing`)
[5] published on the *OpenEnergy Platform* (`http://oep.iks.cs.ovgu.de/dataedit/view/grid`, tables with tag *eTraGo*, dataset version 0.2.11)

its application is that none of the data is published under a license, which makes any further publication illegal by German law. Thus, this data cannot be applied for research that is committed to open science. On the other hand, there are open source approaches to model the German transmission grid (e.g. SciGRID, GridKit, OpenGridMap, osmTGmod). These are typically based on crowd-sourced OpenStreetMap data [1]. All OSM-based models use deterministic or heuristic methods to abstract relevant power topology data sets. Electrical properties have to be calculated and assigned via assumptions to complete the geographical information gathered from raw OSM data (see also Section 3).

The OSM power data for the EHV level covers approximately 95 % of the German transmission grid as far as the total length of the lines is concerned. This has proven to be a sufficient topological coverage [1] and has been applied in several academic works [3, 9].

The OSM objects relevant for the topological modelling of a power grid are *ways* and *relations*. *Ways* correspond to physical power lines mapped in OSM and consist of a list of the *nodes* which are connected and attributes which can be observed such as the number of cables and wires. *Relations* can contain a number of *ways* and additional attributes. They are used in OSM to depict electrical power circuits, which might not be obvious from visual inspection of a line, but require a much more in-depth analysis. Relations provide very accurate information relevant for power flow modelling if they are correctly mapped. However, the OSM coverage of relations decreases below 220 kV. Since, in contrast to most other power flow approaches, our model is not limited to the EHV level, but aims to represent all voltage levels (see Figure 1, OSM ways have to be considered for depicting the HV level. In our definition, the HV level consists of the voltage levels 110 kV and 60 kV. For legibility and the fact that there are less than 30 60 kV lines in Germany, we use the terms HV level and 110 kV synonymously in the following. The data on *ways* in the HV-level seems to be sufficiently representative. Thus, the part of the grid model presented in this contribution intends to cover the exact topology of the voltage levels 110 kV and above. At the same time, the low and medium voltage (MV) levels are adressed in [8], which requires a different approach due to the lower coverage of the downstream medium and low voltage levels, which is mainly caused by the higher share of underground cables not mapped in OSM (see Table 1).

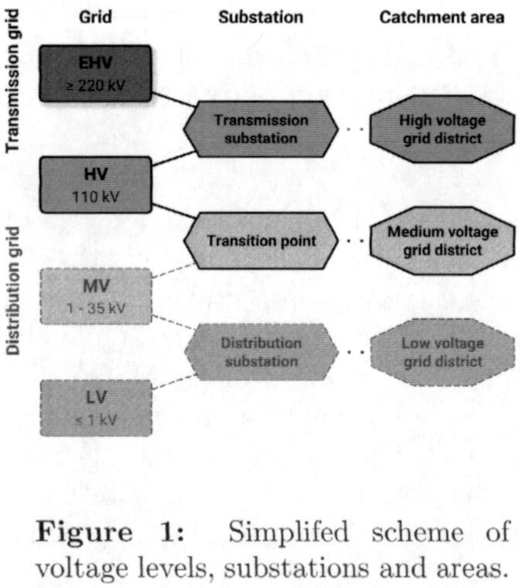

Figure 1: Simplifed scheme of voltage levels, substations and areas. Based on [10]. Figure by Lukas Wienholt / CC BY SA 4.0

Voltage level	OSM relations	OSM ways	Integrated in grid models
380 kV	good	good	yes
220 kV	good	good	yes
110 kV	poor	good	typically not
60 kV	poor	good	typically not
$\leq$ 35 kV	none	poor	no

Table 1: Voltage levels relevant in the German transmission grid, their respective coverage in the OSM database by *relations* and *ways*, and their role in state-of-the-art power flow models. In this context the distinction between *good* and *poor* can be regarded as the adequacy to use the data in power flow simulations.

The open source grid model SciGRID only uses the OSM data type *relation* in order to abstract power circuits [1]. Such an approach can be applied when modeling the EHV grid, due to the weaker mapping quality (see Table 1) it is, however, by far too incomplete to be applied to lower voltage levels. For instance, applying the SciGRID model to the 110 kV voltage level resulted in a total power circuit length of 13, 256 km, whereas the number provided by the Bundesnetzagentur accounts for 96, 658 km [11]. Therefore, our grid model was based on the more heuristic abstraction tool osmTGmod.As a key feature, osmTGmod uses both OSM *relation* and OSM *way* for the construction of a grid model: while the information gathered from *relations* is considered to be very accurate, it is used in the first place. Afterwards, information that is not covered by *relations* is added throught the consideration of *ways*. Due to the lower mapping quality and the existence of 110 kV underground cables in urban areas, there are substations or subgrids initially not connected to the rest of the grid. In our model, transition points are defined as substations, which transfer power flows between HV and MV level (see Figure 1). Since load and generation for power flow simulations are allocated to these transition points, their connection to the main grid as well as the connection of subgrids was essential. Finally, by automatically assigning transformers and electrical properties (see Section 3), the resulting grid model of osmTGmod is ready for power flow simulations without the necessity of further major adaptions [1].

We extended the grid model by an abstraction of its electrical neighbours considering the German electricity grid being embedded into the European transmission grid. Therefore, we placed buses and transformers for different voltage levels at the center of each of the relevant countries. All existing border crossing lines were then connected to these newly introduced buses with respect to their voltage level. The electrical properties of these power lines were determined based on information extracted from the ENTSO-E transmission system map [12] and standard overhead line parameters defined in Section 3.

Applying an OSM data set from October 1st, 2016, the abstraction with osmTGmod leads to a topology model with 11,294 buses and 19,605 branches. Detailed numbers per parameter can be found in Table 2.

Parameter	**Quantity**	
Buses		11,294
thereof	*osmTGmod*	*open_eGo filter*
Substations	4,538	3,702
Joints	6,756	7,592
Branches		19,605
thereof		
Overhead lines		18,221
Underground cables		869
Transformers		515
DC links		5

Table 2: Characteristic results for the grid topology model. The buses resulting from the osmTGmod abstraction have been further filtered for a robust allocation procedure. Buses consist of substations and joints. While substations have central functions, e.g. transformers between differenct voltage levels or allocated load and generation, joints are only nodes linking branches.

The distinction between the substations by osmTGmod abstraction and those filtered for the developed model was necessary for a comprehensive allocation of generation and demand. The mentioned number of 3,702 distinct substations consists of 3,608 HV-MV substations and 94 sole EHV substations. A validation of these transition points and their surrounding grid districts was carried out in [10]. It was shown that the method to identify these areas based on OSM and to allocate generation and demand is applicable and leads to realistic results. Since there is

IOP Conf. Series: Journal of Physics: Conf. Series **977** (2018) 012003 doi:10.1088/1742-6596/977/1/012003

no reference grid model of the HV grid in Germany, the topology can only be validated based on certain DSO areas. A validation was performed based on visual comparision of our topology model to the grid map published by the Southern German DSO LVN [13]. It became evident that the eGo grid model covered every single line at the HV and EHV-levels included in the maps provided by the operator.

3. Assumptions on electrical properties of the grid topology

Electrical proporties had to be assumed and assigned to the generated grid topology in order to perform power flow analyses. Assumptions on lines and transformers were particularly relevant in this context. The assignment of electrical parameters to the topology is part of the tool osmTGmod.

Within each voltage level one standard overhead line and one underground cable was defined representing typically used assets in Germany (see Table 3).

V_{nom} in kV	type	S_{nom} in MVA	R' in Ω/km	L' in mH/km	C' in nF/km
110	line	260	0.109	1.2	9.5
110	cable	280	0.0177	0.3	250
220	line	520	0.109	1	11
220	cable	550	0.0176	0.3	210
380	line	1790	0.028	0.8	14
380	cable	925	0.0175	0.3	180

Table 3: Electrical parameters of standard overhead lines and underground cables for the EHV and HV level. Source: based on [14]

In order to assign and adjust the adequate standard information to the OSM based power lines, information about voltage level (OSM-key *voltage*, V_{nom}), length (implicitly derived from the geometry, l_{osm}) and number of conductors (OSM-key *cables*, $cables_{osm}$) of the OSM topology was used (see also [15, 4, 1]). The number of circuits (or systems) of a route between two grid buses is calculated by dividing the OSM-key *cables* by three ($n_{circuits} = \frac{cables_{osm}}{3}$). This assumption was made because of the exclusive existence of a three-phase AC system in the public grid of Germany and Europe [16]. The resistance R, the reactance X, the capacitance C and the thermal limit apparent power S_{nom} for all lines were then calculated as stated in Equations 1 - 4 (see [1]). Within these equations the standard per unit length values (R'_{lit}, L'_{lit}, C'_{lit}) as well as the circuit-specific $S_{nom,lit}$, which are based on Table 3, were utilized. Consequently, the angular frequency ($\omega = 2 \cdot \pi \cdot f$) corresponding to $f = 50Hz$ was considered.

$$R_{line} = \frac{l_{osm} \cdot R'_{lit}}{n_{circuits}} \tag{1}$$

$$X_{line} = \frac{l_{osm} \cdot L'_{lit} \cdot \omega}{n_{circuits}} \tag{2}$$

$$C_{line} = l_{osm} \cdot C'_{lit} \cdot \omega \cdot n_{circuits} \tag{3}$$

$$S_{nom,line} = S_{nom,lit} \cdot n_{circuits} \tag{4}$$

Modelling all possible interconnections between the voltage levels, three representative transformers were defined (see Table 4). Whenever lines of different voltage were connected at one OSM bus, two buses for each voltage were created and connected by a transformer.

IOP Conf. Series: Journal of Physics: Conf. Series **977** (2018) 012003 doi:10.1088/1742-6596/977/1/012003

The nominal capacity of the transformer was calculated by the minimal sum of $S_{nom,line}$ of all power lines of the same voltage. Thus, the transformer is not expected to be the bottleneck in power flow simulations [15]. The calculated nominal capacity defined the number of transformers installed with respect to the specific S_{nom} in Table 4. The impedance Z of each transformer which was assumed to be equal to the reactance X [17] was calculated as defined in Equation 5. The utilized standard values for the relative short circuit voltage ($v_{sc,lit}$) as well as the higher (V_a) and lower nominal voltages (V_b) were based on Table 4.

$$Z_{tran} = X_{tran} = v_{sc,lit} \cdot \frac{V_a^2}{S_{nom,lit}} \tag{5}$$

S_{nom} in MVA	V_a in kV	V_b in kV	v_{sc} in %
1,000	380	220	13.5
300	380	110	14
200	220	110	12

Table 4: Electrical parameters of standard transformers within the EHV and HV level. Source: based on [18]

The assumptions on the definition of different control strategies at the buses were based on [19]. As usual, the bus connected to the largest generation throughout the year was selected as the slack bus. The slack generation has to be realized by a large and flexible power plant. Large power plants ($P_{nom} \geq 50\,MW$) were defined as PV generator characterizing their connecting buses as PV buses. PQ nodes were buses, which loads or small power plants ($P_{nom} < 50\,MW$) are connected to. Joints without generation or consumption were also modelled as PQ buses.

4. Characterisation of status quo and future scenarios

Three scenarios were defined and used for the intended power flow simulations. Apart from one status quo scenario representing the German electrical energy system in 2015, two future scenarios were defined employing exogenous assumptions. For Germany, the installed generation capacities of the status quo scenario were taken from the power plant list of the Open Power System Data project [20, 21] (State: 01-01-2016). Whereas the 2035 scenario is based on publicly available information and methods of the *Netzentwicklungsplan (NEP) Strom 2025, erster Entwurf* [22]. Out of several NEP scenarios, the so-called "B1-2035" was chosen; it is characterized by a high RE expansion and an increased share of natural gas [22]. The third scenario pictures a future electrical energy system powered to 100% from RE and is mainly based on the *100% RES* scenario of the *e-Highway2050 - Modular Development Plan of the Pan-European Transmission System 2050* [23]. In order to build a 100% energy system in Germany 13 GW of gas fired power plants were removed (see: [24] and [25]). Table 5 shows the scenario specifications and significant characteristics of all three scenarios.

	status quo	**NEP 2035**	**eGo 100%**
Share of RES in installed capacity	46.46%	70.02%	100.00%
Net electricity consumption (TWh)	506.00	506.00	506.00
Annual peak load (GW)	87.01	87.01	87.01
Share of renewable energy in el. consumption[6]	27.31%	65.80%	100.00%

Table 5: Characterization of scenarios by key parameters.

IOP Conf. Series: Journal of Physics: Conf. Series **977** (2018) 012003 doi:10.1088/1742-6596/977/1/012003

A few general characteristics were assumed to be the same in all three scenarios. This includes the year used as a source for data on weather and demand, which is the year 2011 (compare [22]). In terms of weather, the year 2011 can be regarded as a moderate wind feed-in year in Germany [27]. Due to the weather-dependency of electrical demand, one and the same year was chosen as the source for data on weather and demand. Different to the original sources of the scenario parameters, all three scenarios were assumed to have the same load conditions (see Table 5).

We used historical hourly reanalysed weather data from CoastDat-2 [28] to generate temporal and spatial resolution input data for the renewable generation time series (see Section 6.2). The spatial focus of all three scenarios is Germany and its electrical neighbours (AT, CH, CZ, DK, FR, LU, NL, NO, PL, SE). The definition of installed capacities in the neighbouring power systems for the status quo and NEP 2035 scenario was taken from the *Scenario Outlook & Adequacy Forecast (SO&AF) scenario B of 2014* at the reference time point of the annual peak load [29, 30].

The marginal costs are central values to predefine the market decision on operation by the merit order in the Linear Optimal Power Flow and the market simulation of renpassG!S [31]. The marginal costs were calculated as described in Equation 6.

$$C_{marginal} = C_{operating} + C_{fuel} + C_{CO_2 emissions} \tag{6}$$

The CO_2 costs for the status quo scenario are set to 5.91 EUR/t_{CO_2} (EEX Germany mean value 2014) and assumed to be 31.00 EUR/t_{CO_2} for NEP 2035 [22] and 62.05 EUR/t_{CO_2} for the eGo 100% scenario [32]). All other assumptions and calculations in order to simulate the three scenarios are fully documented and can be found in [25].

Table 6 displays the resulting installed electrical capacities in Germany and marginal cost assumptions for each scenario divided by the different technologies and fuels.

[6] Statistic and report values (See: BMWi 2017, sheet 20 year 2014 [26], NEP 2025 [22], own calculation for eGo 100%.)

IOP Conf. Series: Journal of Physics: Conf. Series **977** (2018) 012003 doi:10.1088/1742-6596/977/1/012003

Technology	**status quo**		**NEP 2035**		**eGo 100%**	
	GW	EUR/MWh	GW	EUR/MWh	GW	EUR/MWh
Nuclear energy	12.0	4.68	0.0	5.48	0.0	–
Lignite	23.3	10.78	9.1	17.64	0.0	–
Hard coal	31.5	14.95	11.0	24.79	0.0	–
Natural gas	27.9	32.30	33.5	41.93	0.0	56.05
Oil	4.5	41.02	0.5	68.86	0.0	–
Waste	1.7	31.65	0.0	39.93	0.0	–
CHP < 10 MW [7]	0.0	23.96	8.2	31.11	0.0	31.63
Other conventional generation (mixed fuels)	2.6	31.65	2.4	39.93	0.0	–
Total conventional generation	**103.5**		**64.7**		**0.0**	
Wind onshore	41.3		88.8		98.4	
Wind offshore	5.6		18.5		27.0	
Photovoltaic	38.5		59.9		97.8	
Biomass	7.2		8.4		27.8	
Hydro power	3.9		4.2		4.3	
Other renewable generation	1.4		1.2		2.2	
Total renewable generation	**97.9**		**181.0**		**257.5**	
Pump storage	9.3		12.7		12.8	
Total generation	**210.7**		**258.4**		**270.3**	

Table 6: Installed generation capacities in GW for Germany and marginal costs for conventional generation in $EURO_{2014}/MWh$, divided by scenario and technology/fuel. Source: [25]

Apart from the exogenous scenario parameters already mentioned, standard power plant parameters such as onshore wind power plant size, height or efficiency factors were defined for the respective generator types and subtypes in Section 6.2.

5. Allocation of loads and generators

The following section describes the spatial allocation of generators and loads to the predefined grid topology. Since information on generation and demand affecting the grid is compulsory, generators and loads needed to be assigned to the respective grid level and the relevant substation within this grid level. Geographic catchment areas, which differ according to the considered grid level, were the basis for the spatial assignment to substations. The catchment areas of EHV substations were defined by a Voronoi partition, a well-known approach used in different publications [3, 33]. In [10], this approach is expanded using administrative borders as an additional input for the representation of the underlying MV grids.

5.1. Load allocation

Data on the electricity consumption throughout Germany for the reference year 2011 is available on the level of the federal states [34]. This spatial resolution does not meet the requirements of the intended powerflow simulations. Therefore, an allocation of the German electricity demand was implemented. The approach used to distribute and allocate loads with a high spatial resolution is described in detail by [10].

[7] For the status quo and eGo 100% scenario small CHP units are included in the accounts of biomass and natural gas. The marginal costs of CHP are referred to the costs for biomass.

IOP Conf. Series: Journal of Physics: Conf. Series **977** (2018) 012003 doi:10.1088/1742-6596/977/1/012003

In this approach, information about the spatial distribution of population, the gross value added and the industrial and retail area functioned as the base for a spatial allocation of electricity demand. The assignment of distributed loads to their relevant bus was achieved using grid districts as a representation of the underlying MV grid for loads connected to the lower grid levels. Voronoi cells were applied for the allocation of industrial large scale consumers directly connected to the EHV level.

5.2. Generation allocation

To link data on the electricity generation with our grid topology, we derived information on power plants from two different registries for the status quo scenario - the conventional power plant list [20] containing large, mainly fossil power plants $> 10\ MW$ and a renewable power plants list [21]. All records in these registries are georeferenced. In the renewables power plant registry the minimum geographical accuracy is the maximum distance of a post code area. The conventional power plant data were manually georeferenced within the OPSD project using their addresses. We assigned generators to a voltage level, followed by a geographical allocation to the respective substation based on its catchment area. In our model, all offshore wind parks were connected to the EHV level. Information on the net connection points of present and future offshore wind parks are available in [35] and used to assign wind offshore units to their respective substation in the developed grid model. For future wind parks, which are according to [35] connected to planned substations not represented in the grid model, an existing nearby substation was chosen.

For the NEP 2035 scenario, information on conventional power plants were extracted from [36] and have automatically or manually been georeferenced. From this source only the large hydro and pumped storage plants are included in the eGo 100% RES scenario.

Beside conventional power plants individually listed in the aforementioned source, aggregated future capacities need to be distributed geographically to meet the need for a high spatial resolution. For the NEP 2035 the expansion of installed capacity per technology and federal state in Germany served as the starting point. Whereas the allocation for the 100% scenario was based on the assumption of installed capacity on country level. The methods and assumption of the spatial allocation differ depending on the type of RE generation.

For geothermal power plants no changes were assumed for future scenarios. The development of biomass, small gas and hydro power plants was expected to increase proportionally to its specific value of installed capacity based on the status quo in 2014. Due to the high development of small CHP plants within the NEP scenario, a geographical allocation method of the additional plants was developed and used [37].

The location of wind offshore units in the NEP 2035 scenario was taken from the *Netzentwicklungsplan Offshore 2025 (O-NEP)* [35, 38]. New wind onshore and photovoltaic systems were as well assumed to see a proportional development starting from the status quo. These plants were allocated on the municipality level which is sufficient for analyses of the HV and EHV grid. The initial information was the number and size of renewable power plants per municipality. Based on this, representative units were defined for each municipality and were then proportionaly extended in future scenarios. The location of the new power plants is the centroid of the respective municipality. A more detailed allocation process which becomes necessary when addressing low and medium voltage levels is part of [8].

The allocation for the 100% scenario data is based on the same methods and its scenario-specific assumptions of installed capacities per technology. The amount of power plants increases from the status quo scenario to the 100% scenario especially due to the growing number of renewable power plants which affects the complexity of the analyses per scenario.

6. Time series of load and generation

Calculations on the grid model require a suitable characterization of electricity demand and supply at grid nodes. This was achieved by describing demand and supply characteristics with time series of one year, which allows for a detailed analysis of the electricity system with a focus on grid operation. This data must reflect generation and demand of a certain area to which the grid node is associated.

6.1. Demand time series

The aim was to obtain a consistent set of demand data valid across all grid levels in Germany. Numerous approaches exist to model residential electricity load curves [39]. These range from bottom-up modeling of residential load curves considering behaviourial factors [40] to electricity load curves of German residential sector at high resolution [41]. The mentioned approaches are rather problem-specific and do not provide a complete data set for all sectors considered in this study (residential, industry, agricultural and retail). A comprehensive modeling approach for the modeling of electricity demand across grid levels is not available either. This involves an additional challenge as the coincidence of electricity demand varies across grid levels respectively aggregation levels. The shape of a load curve depends on the number of customers represented. The shape of a cumulative load curve of more than 1,000 customers does not change anymore [42].

In Germany, electricity demand characteristics at low-voltage grid level are usually described by standard load profiles (SLP) [43]. This information is provided according to sector (residential, retail, agricultural) and at a temporal resolution of 15-minutes. The data is based on measurements of 332 households. The authors state this is valid as a cumulative load for a number of customers with similar characteristics – without providing any quantity [44]. Willis and Scott argue that already the cumulative demand curve of 100 households "...looks smooth and 'well-behaved.'" [42]. Thus, we assume SLP are valid in a range of 100 to 300 residential customers. Additionally to demand patterns for these three sectors, a pattern for electricity demand of the industrial sector was constructed based on a stairs function [45]. The industrial demand pattern considered peak and off-peak times. During a work week at day-time (6 a.m. to 10 p.m.) the normalized load curve adds up to 0.8. At other times this parameter was set to 0.6.

Spatially highly resolved and sectorally disaggregated annual consumption $C_{j,s}$ (described above) was converted to hourly demand time series $P^{\circ}_{demand,j}$ by applying sectoral demand patterns $P^{\circ}_{SLP,s}(t)$ (see Equation 7).

$$P^{\circ}_{demand,j}(t) = P^{\circ}_{SLP,s}(t) \cdot C_{j,s}, \tag{7}$$

where s refers to the sectors residential, retail, industrial and agricultural and j denotes the substation. We assumed a power factor $\cos\phi$ of 0.95 (inductive) for aggregated loads at the HV substation [46]. Annual consumption, originally determined for each *load area* [10], was aggregated to HV grid nodes. [8]

6.2. Generation time series

For each scenario, generation time series are based on hourly global power market simulation of Germany and its neighbouring countries. The open source model renpassG!S, which is an application of the Open Energy Modelling Framework (oemof) and the oemof library Feedinlib, was applied for this [31, 48, 49]. The oemof library Feedinlib provides time series for photovoltaic systems and wind power plants sorted by standard technical parameters of power plant types and

[8] The implementation is based on *oemof-demandlib* [47] and located in the *data processing* of the *open_eGo* project.

based on the weather data model CoastDat-2 [49, 28]. The calculated time series per country were proportionately allocated to the substations considering the cumulated installed electrical capacity by technology per substation.

The usage of generation time series differs according to the type of grid calculation. The non-linear Newton Raphson Power Flow (PF) and Linear Power Flow (LPF) calculations use given time series per generation type. The Linear Optimal Power Flow (LOPF) on the other hand only uses the weather-dependent time series of wind and solar feed-in. The time series of other generation types are a result of the LOPF being based on the same market principle as the global power market simulation, but at the same time considering network constraints in a DC flow abstraction. Furthermore, the reactive power behaviour (relevant for the PF) was defined by a $\cos(\phi) = 1$ for small power plants ($P_{nom} < 50\,MW$) whereas large power plants being PV generators have a flexible $\cos(\phi)$ [19].

7. Application of the data model in power flow calculations

The data processing methods described in Sections 2 to 6 provide the necessary basis for power flow simulations and optimizations in a high temporal and spatial resolution. Exemplary results of these analyses are presented in this Section. The open source tool *PyPSA*[9], which was integrated into the *eTraGo* app[10], was the basis for the simulations and allows several power flow methods. Besides standard PF and LPF simulation methods a LOPF method optimizes power plant dispatch considering network constraints. The grid data model is suitable for all of the methods mentioned above, the analyses and results presented in this contribution are, however, based on LOPF. The LOPF additionally allows investive optimization of capacity extension, which will be a major part of future open_eGo modelling, but is not considered in this work. Out of the three scenarios defined (see Section 4), the focus will be on the status quo and NEP 2035 scenarios in the following.

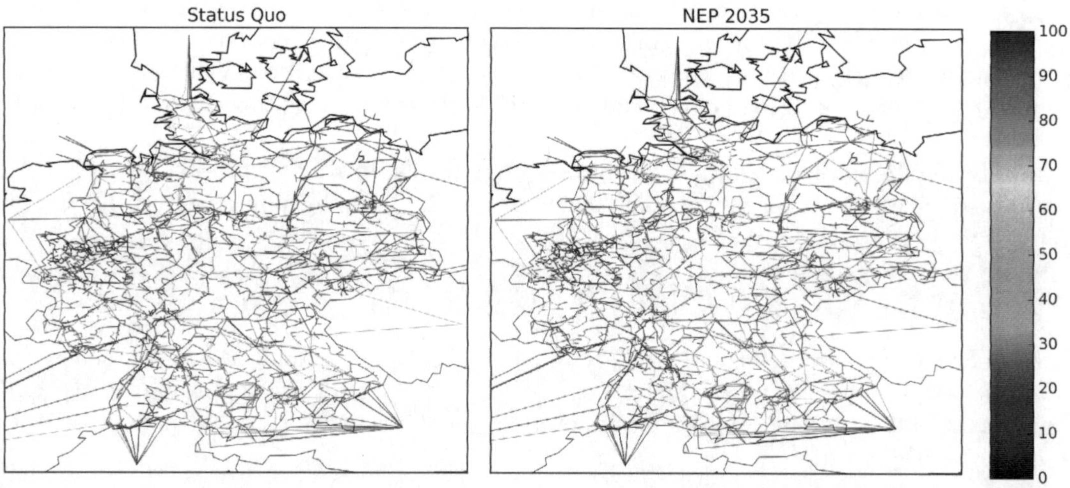

Figure 2: Relative line loading in % for the status quo and NEP 2035 scenarios at the hour of the lowest overall residual load throughout the year (7th of April, 2 p.m.). The three voltage levels (380, 220, 110 kV) are displayed without explicit differentiation. Sweden, the Czech Republic and France are not displayed with their centroids for better visualization. Figure by Ulf Philipp Müller / CC BY SA 4.0

[9] `www.pypsa.org`

[10] *eTraGo* (`https://github.com/openego/eTraGo`) and *ego.io* (`https://github.com/openego/ego.io`) were primarily used as an interface between the data model's database and PyPSA.

The exemplary results are based on a 24 hour-optimization of the day with the lowest overall residual load of the modelled year (7th of April). Therefore the share of renewable power feed-in is at its highest [50]. Estimating approximately the n-1 security requirements the maximal possible branch capacities were globally derated to 70% of the individual thermal limits [51].

For the sake of problem feasibility[11] at each bus load could be shedded considering a value of lost load of 10,000 Euro/MWh [52]. The reasons for the feasiblity problems are currently being researched. In this context, the high number of lines and buses represents a challenge. Analyses showed that only at a small number of buses (0.6 %) located in the 110 kV level load shedding occurs. These critical spots are located mostly in big cities such as Berlin and Munich where the degree of underground cable is rather high and therefore OSM mapping quality low. The poor mapping quality (concering the number of cables for example) displays the most probable explanation for these few local feasibility problems. Moreover the general assumptions on electrical parameters (concerning the nominal capacities, e.g the number of wires) or weaknesses in the allocation of loads and generators which could for instance result in too much load allocated to the 110 kV grid level are considered as possible sources.

The relative line loading of one hour for the status quo and NEP 2035 scenario can be examined in Figure 2. In the status quo scenario 392 power lines are loaded with more than 50 % compared to their thermal limits. In the future scenario, 747 power lines exceed this relative value. In the status quo scenario particularly the 110 kV lines in Schleswig-Holstein showed high usage rates. In contrast in the 2035 scenario the line loading increases in multiple regions whereas it decreases on many of the 110 kV lines in Schleswig-Holstein. The nearly omnipresent abundance of renewable power plants (particularly wind power plants) and supraregional grid restrictions (status quo grid was applied for 2035 scenario) cause these characteristics including substantial curtailment of RE. The modelled transformers do not show any critical congestions.

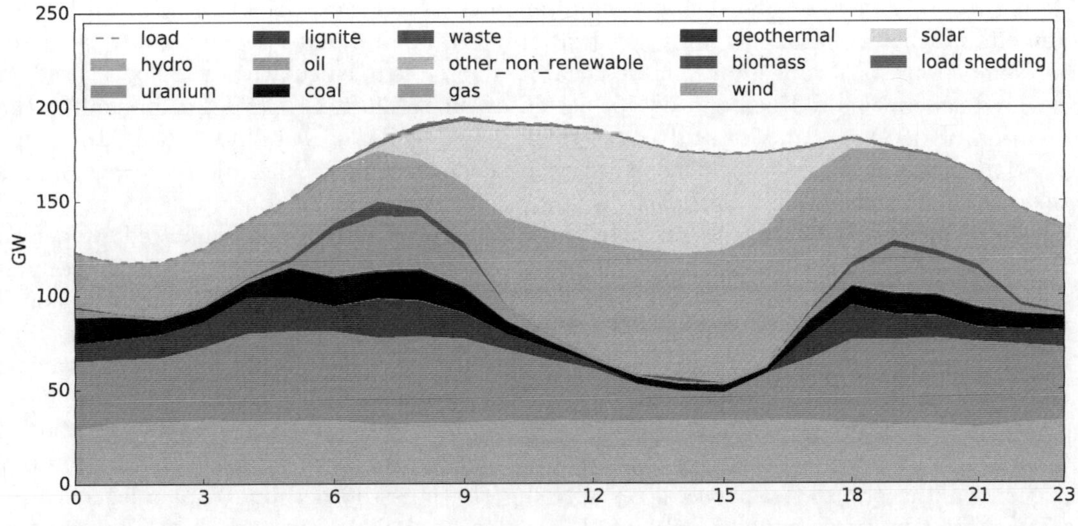

Figure 3: Stacked generation and load for the scenario NEP 2035 on the 7th of April(0 a.m. - 23 p.m.) aggregated for Germany and its electrical neighbouring countries. Figure by Ulf Philipp Müller / CC BY SA 4.0

The data model allows the calculation of 8,760 hours for each scenario year. In Figure 3, the stacked generation for one day within the 2035 scenario year is displayed. At 2 p.m. the

[11] glpk (`https://www.gnu.org/software/glpk/`) and gurobi (`http://www.gurobi.com/` are used as solvers. These particular results were generated using gurobi due to lower computing times.

generation mix corresponds to the line loading shown on the right side of Figure 2. At this hour almost the entire load of Germany and its electrical neighbouring countries is covered by renewable generation. The nuclear power generation is originated from power plants located outside of Germany (e.g. France). The lignite and hard coal fired power plants can hardly operate during this midday period. During this day 36 % of the available solar and wind energy are curtailed. Due to feasibility problems, load shedding can be observed during daytime when demand is high. Its share in total demand throughout this day is with 0.4 % relatively low.

The results presented in this Section based on an exemplary day indicate that the open source grid data model allows extensive analyses of power plant dispatch under consideration of grid restrictions. The existing uncertainties regarding the problem feasibility and general complexity are essential parts of current and future research.

8. Conclusion and Outlook

An approach to create a model integrating the existing German EHV and HV grid was developed and implemented. The integration of the 110 kV grid displays an innovation in the field of open data grid models for Germany. At the same time the electrically neighbouring countries were abstracted considering their electrical and dispatch influence. The result is a plausible grid topology model with a high number of buses (11,294) and branches (19,605), which is a suitable basis for state-of-the-art power flow methods (PF, LPF and LOPF).

Completing the developed model, electricity demand and generation were defined for three different scenarios in order to derive results for the present, the mid and long-term. The allocation and aggregation of demand and generation units to each HV-MV substation (3,608) considering an interface to the MV and LV levels represents an adequate degree of spatial resolution. Per scenario hourly values were determined for each unit or aggregate over the course of one year. This allows research on dispatch and redispatch throughout an entire year considering realistic weather and demand simultaneities. Moreover, dispatch optimization for storage units which are characterized by substantial intertemporal dependencies can be realized.

The exemplary results in Section 7 showed that it is possible to calculate power flows based on the allocated generation and demand over a period of time. The spatial resolution enables the identification of decentral and central congestions of particular power lines within the same optimization problem. The weather-dependent behaviour of RE generation as well as the marginal cost driven dispatch optimization was shown to work conceivably.

The entire model (grid, generation, demand) has to be tested and validated in detail. The methodologies presented show high potential and produce convincing exemplary results. However, the results also disclose some difficulties. Minor load shedding at few buses was used to reach a feasible optimization problem. Recently grid operators' publications[12] were used to validate and vary line assumptions within the EHV level[53]. Therefore modifications can be deduced in the near future. In this context it is also discussed to utilize the OSM-tag *wires*, which can be applied on the EHV and HV level. As the tag *wires* represents the number of conductors per phase for a power line, an integration of this tag may lead to more detailed information on the transfer capacities. Neglecting the cross-border line capacities between the neighboring countries influences the model results (especially because loop flows are not considered). The implementation of such interconnections can be realized in the near future expecting an enhancement for the model. The further development is a continuous process, which is facilitated by the applied open source standards. All developed methodologies are reproducible and in this sense meet scientific standards to a high extend.

Based on the presented grid model future contributions will focus on the optimization of eventual investment decisions concerning storage and grid expansion. The spatial and temporal

[12] e.g. for transmission system operator *Tennet TSO* see `https://www.tennettso.de/site/Transparenz/veroeffentlichungen/statisches-netzmodell/statisches-netzmodell`

complexity and a high number of optimization variables lead to immense computation times when applying deterministic optimization methods. Accordingly complexity reduction methods (e.g. snapshot clustering) are currently being developed.

Acknowledgments
The authors would like to thank the *German Federal Ministry for Economic Affairs and Energy* for funding the research project *open_eGo*. Funding code: 0325881A-E. Additionally, the authors thank Malte Scharf for the creation of *osmTGmod* as well as Tom Brown and Jonas Hoersch of the *Frankfurt Institute for Advanced Studies* for the development of *PyPSA*.

References
[1] Medjroubi W, Müller U P, Scharf M, Matke C and Kleinhans D 2017 OpenData in Power Grid Modelling: New Approaches Towards Transparent Grid Models *Energy Reports* **3** 14–21
[2] Büchner J, Katzfey J, Floercken O D, Moser A, Schuster H, Dierkes S, van Leeuwen T, Verheggen L, Uslar M and van Amelsvoort M Verteilernetzstudie für Deutschland (Verteilernetzstudie): Forschungsprojekt Nr. 44/12: Abschlussbericht URL http://www.bmwi.de/BMWi/Redaktion/PDF/Publikationen/Studien/verteilernetzstudie,property=pdf,bereich=bmwi2012,sprache=de,rwb=true.pdf
[3] van Leeuwen T, Dierkes S, Verheggen L, Schuster H, Köhne F and Moser A 2014 Ermittlung von Transitflüssen im Hochspannungsnetz durch mehrere Verknüpfungspunkte mit dem Übertragungsnetz *13. Symposium Energieinnovation, 12.-14.2.2014, Graz/Austria* URL https://www.tugraz.at/fileadmin/user_upload/Events/Eninnov2014/files/kf/KF_vanLeeuwen.pdf
[4] Müller U P, Cussmann I, Wingenbach C and Wendiggensen J 2016 AC Power Flow Simulations within an Open Data Model of a High Voltage Grid *Advances and New Trends in Envrionmental Informatics* ed Wohlgemuth V, Fuchs-Kittowski F and Wittmann J (Springer International Publishing Switzerland) pp 181–191
[5] Rippel K M, Wiede T, Meinecke M and König R 2017 Netzentwicklungsplan Strom 2030, Version 2017 – Erster Entwurf der Übertragungsnetzbetreiber URL https://www.netzentwicklungsplan.de/de/netzentwicklungsplaene/netzentwicklungsplaene-2030
[6] Agricola A C, Höflich B, Richard P, Völker J, Rehtanz C, Greve M, Gwisdorf B, Kays J, Noll T, Schwippe J, Seack A, Teuwsen J, Brunekreeft G, Meyer R and Liebert V Ausbau- und Innovationsbedarf der Stromverteilnetze in Deutschland bis 2030 (kurz: dena-Verteilnetzstudie): Endbericht URL http://www.dena.de/fileadmin/user_upload/Projekte/Energiesysteme/Dokumente/denaVNS_Abschlussbericht.pdf
[7] Bundesministerium für Wirtschaft und Energie 2017 Gesetz für den Ausbau erneuerbarer Energien (Erneuerbare-Energien-Gesetz - EEG 2017)
[8] Amme J, Pleßmann G, Bühler J, Hülk L, Kötter E and Schwaegerl P 2017 The eGo grid model: An open-source and open-data based synthetic medium-voltage grid model for distribution power supply systems *SciGRID International Conference on Power Grid Modelling* (submitted)
[9] Hinz F and Möst D 2016 The effects of reactive power provision from the distribution grid on redispatch cost *2016 13th International Conference on the European Energy Market (EEM)* pp 1–6
[10] Huelk L, Wienholt L, Cussmann I, Mueller U P, Matke C and Koetter E 2017 Allocation of annual electricity consumption and power generation capacities across multi voltage levels in a high spatial resolution *International Journal of Sustainable Energy Planning and Management* (submitted)
[11] Bundesnetzagentur and Bundeskartellamt 2016 Monitoringbericht 2016 URL https://www.bundesnetzagentur.de/DE/Sachgebiete/ElektrizitaetundGas/Unternehmen_Institutionen/DatenaustauschundMonitoring/Monitoring/Monitoringberichte/Monitoring_Berichte_node.html
[12] ENTSO-E 2016 ENTSO-E Transmission System Map 2016 URL https://www.entsoe.eu/map/Pages/default.aspx
[13] LEW Verteilnetz GmbH 2014 LVN Netzplan URL https://www.lew-verteilnetz.de/media/5655/netzplan.pdf
[14] Brakelmann H 2004 Netzverstärkungs-Trassen zur Übertragung von Windenergie: Freileitung oder Kabel? URL http://www.ieea.uni-duisburg.de/~brakelmann/Freileitung_Kabel.pdf
[15] Scharf M 2015 *Entwicklung eines Modells des deutschen Übertragungsnetzes auf Basis der offenen Geodatenbank OpenStreetMap* Bachelor-thesis University of Applied Sciences Flensburg
[16] Schwab A 2012 *Elektroenergiesysteme: Erzeugung, Transport, Übertragung und Verteilung elektrischer Energie* 3rd ed (Springer-Verlag)

[17] Flosdorff R and Hilgarth G 2005 *Elektrische Energieverteilung (Leitfaden der Elektrotechnik)* vol 9 (B. B. Teubner Verlag)
[18] Oeding D and Oswald B R 2011 *Elektrische Kraftwerke und Netze* (Springer)
[19] Rendel T 2015 *Erweiterung und Plausibilisierung eines Modells für die integrierte Simulation des europäischen Verbundnetzes und Strommarktes* Ph.D. thesis Gottfried Wilhelm Leibniz Universität Hannover
[20] Gerbaulet C and Kunz F 2016 Data provided by Open Power System Data - Data Package Conventional power plants, version 2016-10-27. Primary data from BNetzA Kraftwerksliste, Umweltbundesamt Datenbank Kraftwerke in Deutschland URL `http://open-power-system-data.org/`
[21] Bunke W D 2016 Data provided by Open Power System Data - Data Package Renewable power plants, early version 2016-02-10. Primary data from BNetzA, BNetzA_PV, TransnetBW, TenneT, Amprion, 50Hertz, Netztransparenz.de, Postleitzahlen Deutschland, Energinet.dk, Energistyrelsen, GeoNames, French Ministery of the Environment, Energy and the Sea, OpenDataSoft, Urzad Regulacji Energetyki (URE) URL `http://open-power-system-data.org/`
[22] 50Hertz Transmission GmbH, Amprion GmbH, TenneT TSO GmbH, TransnetBW GmbH (ÜNB) 2015 Netzentwicklungsplan Strom 2025, Version 2015 – Erster Entwurf der Übertragungsnetzbetreiber URL `https://www.netzentwicklungsplan.de/sites/default/files/paragraphs-files/NEP_2025_1_Entwurf_Teil1_0.pdf`
[23] e Highway2050 2015 e-highway 2050 modular development plan of the pan-european transmission system 2050 - database per country URL `http://www.e-highway2050.eu/fileadmin/documents/Results/e-Highway_database_per_country-08022016.xlsx`
[24] Christ, M 2017 *Integration sozial-ökologischer Faktoren in die Energiesystemmodellierung am Beispiel von Entwicklungspfaden für den Windenergieausbau in Deutschland* Ph.D. thesis Europa-Universität Flensburg (unpublished)
[25] Bunke, W D and Söthe, M and Christ, M and Kaldemeyer, C 2017 (Fl)ensburg (En)ergy (S)cenarios - open_eGo Scenarios for 2014/2035/2050 URL `https://osf.io/bpf36/`
[26] Bundesministerium für Wirtschaft und Energie BMWi 2017 Gesamtausgabe der Energiedaten - Datensammlung des BMWi (Sheet 20) URL `http://www.bmwi.de/Redaktion/DE/Artikel/Energie/energiedaten-gesamtausgabe.html`
[27] 50Hertz Transmission GmbH, Amprion GmbH, TenneT TSO GmbH, TransnetBW GmbH (ÜNB) 2015 NEP 2025 - Methodik der Modellierung wetterabhängiger erneuerbarer Energien (zu Kapitel 2.2.2) URL `https://www.netzentwicklungsplan.de/de/methodik-der-modellierung-wetterabhaengiger-erneuerbarer-energien-zu-kapitel-222`
[28] Geyer B and Rockel B 2013 coastDat-2 COSMO-CLM Atmospheric Reconstruction. World Data Center for Climate (WDCC) at DKRZ
[29] ENTSO-E 2014 Scenario Outlook & Adequacy Forecast (SO&AF) 2014-2030 Dataset URL `www.entsoe.eu/publications/system-development-reports/adequacy-forecasts/soaf-2014-2030/Pages/default.aspx`
[30] 50Hertz Transmission GmbH, Amprion GmbH, TenneT TSO GmbH, TransnetBW GmbH (ÜNB) 2015 NEP 2025 - Weiterführende Informationen zur Nachbildung des Auslandes (zu Kapitel 2.4.2) URL `https://www.netzentwicklungsplan.de/node/359`
[31] Kaldemeyer C, Söthe M, Hilpert S and Wingenbach C 2017 renpassG!S - A Free and Open Python Tool for Simulating Energy Supply Systems. URL `https://osf.io/kp4mh/`
[32] Nitsch, J and Pregger, T and et al 2012 Langfristszenarien und Strategien für den Ausbau der erneuerbaren Energien in Deutschland bei Berücksichtigung der Entwicklung in Europa und global Tech. rep. DLR, IWES, IFNE URL `http://www.dlr.de/dlr/Portaldata/1/Resources/bilder/portal/portal_2012_1/leitstudie2011_bf.pdf`
[33] Singh A, Willi D, Chokani N and Abhari R S 2014 Optimal power flow analysis of a switzerland's transmission system for long-term capacity planning *Renewable and Sustainable Energy Reviews*
[34] Länderarbeitskreis Energiebilanzen 2015 Energiebilanzen der Länder URL `http://www.lak-energiebilanzen.de/seiten/energiebilanzenLaender.cfm`
[35] 50Hertz Transmission GmbH, Amprion GmbH, TenneT TSO GmbH, TransnetBW GmbH (ÜNB) 2015 Offshore-Netzentwicklungsplan 2025 (O-NEP 2025), Version 2015 – Erster Entwurf der Übertragungsnetzbetreiber URL `https://www.netzentwicklungsplan.de/sites/default/files/paragraphs-files/ONEP_2025_1_Entwurf_Teil1.pdf`
[36] 50Hertz Transmission GmbH and Amprion GmbH and TenneT TSO GmbH and TransnetBW GmbH 2014 Kraftwerksliste zum Entwurf Szenariorahmen NEP/O-NEP 2025 URL `https://www.netzentwicklungsplan.de/sites/default/files/paragraphs-files/nep_2015_kraftwerksliste_entwurf_140430.pdf`

[37] Kropshofer M 2017 *Erstellung und Implementierung einer Methodik zur Verteilung von KWK-Anlagen* Project report Europa-Universität Flensburg

[38] 50Hertz Transmission GmbH, Amprion GmbH, TenneT TSO GmbH, TransnetBW GmbH (ÜNB) 2015 Anhang zum Offshore-Netzentwicklungsplan 2025, Version 2015 – Erster Entwurf der Übertragungsnetzbetreiber URL https://www.netzentwicklungsplan.de/sites/default/files/paragraphs-files/ONEP_2025_1_Entwurf_Teil2.pdf

[39] Grandjean A, Adnot J and Binet G 2012 A review and an analysis of the residential electric load curve models *Renewable and Sustainable Energy Reviews* **16** 6539 – 6565 ISSN 1364-0321 URL http://www.sciencedirect.com/science/article/pii/S1364032112004820

[40] Capasso A, Grattieri W, Lamedica R and Prudenzi A 1994 A bottom-up approach to residential load modeling *IEEE Transactions on Power Systems* **9** 957–964 ISSN 0885-8950

[41] Fischer D, Härtl A and Wille-Haussmann B 2015 Model for electric load profiles with high time resolution for german households *Energy and Buildings* **92** 170 – 179 ISSN 0378-7788 URL http://www.sciencedirect.com/science/article/pii/S0378778815000845

[42] Willis L and Scott W 2000 *Distributed Power Generation: Planning and Evaluation* (Marcel Dekker, inc.)

[43] Hayn M, Bertsch V and Fichtner W 2014 Electricity load profiles in Europe: The importance of household segmentation *Energy Research & Social Science* **3** 30–45 URL http://energyexemplar.com/wp-content/uploads/2014/09/Hayn-2014-Electricity-load-profiles.pdf

[44] Meier H, Fünfgeld C, Adam T and Schiefdecker B 1999 Repräsentative VDEW-Lastprofile Tech. rep. Brandenburgische Technische Universität Cottbus

[45] Schachler B 2014 *Bewertung des Einsatzes von Kraft-Wärme-Kopplungsanlagen hinsichtlich der CO2-Emissionen bei wachsendem Anteil Erneuerbarer Energien* Masterarbeit Technische Universität Berlin

[46] Knorr K, Zimmermann B, Kirchner D, Speckmann M, Spieckermann R, Widdel M, Wunderlich M, Mackensen R, Rohrig K, Steinke F, Wolfrum P, Leveringhaus T, Lager T, Hofmann L, Filzek D, Göbel T, Kusserow B, Nicklaus L and Ritter P 2014 Kombikraftwerk 2 – Abschlussbericht

[47] oemof Developer Group 2016 The oemof demandlib (oemof.demandlib) URL https://doi.org/10.5281/zenodo.438786

[48] Hilpert S, Günther S, Kaldemeyer C, Krien U, Pleßmann G, Wiese F and Wingenbach C 2017 Addressing energy system modelling challenges: The contribution of the open energy modelling framework (oemof) *Preprints* URL http://www.preprints.org/manuscript/201702.0055/v1

[49] Krien U and oemof developing group 2016 feedinlib – time series of pv or wind power plants URL https://github.com/oemof/feedinlib

[50] Schill W P 2014 Residual load, renewable surplus generation and storage requirements in germany *Energy Policy* **73** 65–79

[51] Wiese F, Bökenkamp G, Wingenbach C and Hohmeyer O 2014 An open source energy system simulation model as an instrument for public participation in the development of strategies for a sustainable future *Wiley Interdisciplinary Reviews: Energy and Environment* **3** 490–504 ISSN 20418396 URL http://www.bmwi.de/BMWi/Redaktion/PDF/G/gesetz-fuer-den-ausbau-erneuerbarer-energien,property=pdf,bereich=bmwi2012,sprache=de,rwb=true.pdf

[52] Schröder T and Kuckshinrichs W 2015 Value of lost load: An efficient economic indicator for power supply security? a literature review *Frontiers in Energy Research* **3** 55 ISSN 2296-598X URL http://journal.frontiersin.org/article/10.3389/fenrg.2015.00055

[53] Vespermann M 2017 *Entwicklung einer Methodik zur Betriebsmittel-Parametrisierung im Rahmen eines offenen Stromnetzmodells der Höchst- und Hochspannungsebene Deutschlands* Master thesis Europa-Universität Flensburg

IOP Conf. Series: Journal of Physics: Conf. Series **977** (2018) 012004 doi:10.1088/1742-6596/977/1/012004

Impact of Considering 110 kV Grid Structures on the Congestion Management in the German Transmission Grid

André Hoffrichter, Hans Barrios, Janek Massmann, Bhavasagar Venkataramanachar, Armin Schnettler

Institute for High Voltage Technology, RWTH Aachen University, Germany

E-mail: hoffrichter@ifht.rwth-aachen.de

Abstract. The structural changes in the European energy system lead to an increase of renewable energy sources that are primarily connected to the distribution grid. Hence the stationary analysis of the transmission grid and the regionalization of generation capacities are strongly influenced by subordinate grid structures. To quantify the impact on the congestion management in the German transmission grid, a 110 kV grid model is derived using publicly available data delivered by Open Street Map and integrated into an existing model of the European transmission grid. Power flow and redispatch simulations are performed for three different regionalization methods and grid configurations. The results show a significant impact of the 110 kV system and prove an overestimation of power flows in the transmission grid when neglecting subordinate grids. Thus, the redispatch volume in Germany to dissolve bottlenecks in case of N-1 contingencies decreases by 38 % when considering the 110 kV grid.

1. Introduction

The massive integration of renewable energy sources (RES) and the energy market liberalization have led to strong structural changes in the European power system, which are expected to continue. Already today 76% of the installed capacity of RES in Germany is connected to the low-, medium- and high-voltage level with an increasing tendency [1]. Thus, reversing power flows from the distribution into the transmission grid can be observed. In addition, the concentration of RES far from load centers, e.g. wind power in northern Germany, involves very long transmission distances. Both developments change the typical power flow patterns within the power system and hence require a comprehensive analysis of the utilization patterns of the future power system.

Existing grid models for power flow simulations or tools for network expansion planning usually just focus on a detailed representation of the extra high voltage level (EHV) [2][3][4]. However, subordinate high voltage grids, e.g. the 110 kV level in Germany, are totally neglected or not modelled in detail. Instead grid equivalents representing the residual power flows are applied [5][6]. This however stands in clear contradiction to the described structural changes. As a consequence, pursuing methods for an adequate modelling of subordinate high voltage grids are required, which allow an analysis of the impact of subordinate grid structures on the transmission grid. In addition to an accurate representation of substations and lines, a suitable regionalization method is required to map the generation capacities and loads to the nodes of the derived grid model. Based on the resulting grid model a recommendation for the consideration of subordinate high voltage grids in future studies, e.g. power flow or redispatch simulations, may be given.

2. Methodology

2.1. Grid data basis

Due to the comprehensive data and its suitable format, data delivered by Open Street Map (OSM) is used to develop a model of the German 110 kV network. OSM is a geographic information system that, among others, contains geographical information of the electrical infrastructure. In addition, this data mostly contains information regarding voltage level, frequency and number of circuits, which allows an artificial determination of the electrical parameters [7]. OSM data is categorized in nodes, ways and relations with each type having a unique numeric identifier. In addition every node has a unique geographical identifier indicating if it represents, for example, an electrical tower or an arbitrary point on the fence of a substation. Ways are comprised of more than one node and can be classified as a closed-way which represents a substation, generation plant or other electrical infrastructure facilities. An open-way however represents a route comprising of one or more systems of an overhead line (OHL) or an underground cable. This differentiation is directly provided in each way as an attribute-tag with the type of infrastructure it represents. Every relation is comprised of one or more ways and nodes and may represent a substation, generation plant or route.

2.2. Modelling of 110 kV grid structures

Since OSM is an open source project and available information is maintained by volunteers, methods are required to process and supplement data as well as check its plausibility. Therefore, an iterative process is established that successively adjusts existing and complements missing information to derive a node-branch model of the 110 kV grid in Germany. The methodology is implemented in Matlab® and classified into the steps pre-processing, main-processing and post-processing. Since ways and relations not only connect substations, but instead are connected to each other on nodes (towers), the main aspect of the process is to identify how different routes are connected to each other and handle them accordingly. Connections of different routes on towers are classified into connection types based on the number of routes meeting on a tower. Rulesets are defined for each connection type to enable the handling of such multiple routes on towers. The rulesets contain instructions to handle routes by splitting them into individual circuits or generate tripod structures. Figure 1 exemplarily shows three connection types of different routes that have various number of conductors of the three-phase system and associated rulesets to handle them. In total, nearly 350 different connection types are identified within the German 110 kV grid and nearly 20 rulesets are developed to define the galvanic connection of the conductors. The rulesets are used in the pre-processing and main-processing steps to validate and modify the topology.

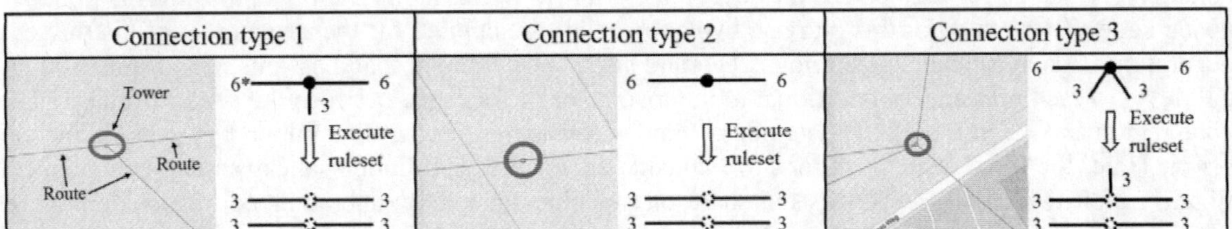

Figure 1. Handling of various connection types for three given examples.

The defined ruleset is used to handle missing or (rare occurring) implausible electrical information of routes in the pre-processing step. Missing information of routes is assumed after analysing the information of all other routes associated with it. This procedure is performed iteratively for voltage, frequency and number of conductor information of all routes in the entire grid. Implausible routes (e.g. connections of lines of different voltage levels or connections of routes where, due to the number of conductors, no technically valid connection can be defined for) are identified and their information is

IOP Conf. Series: Journal of Physics: Conf. Series **977** (2018) 012004 doi:10.1088/1742-6596/977/1/012004

changed depending on the connection type and its associated rule. Subsequently, routes that do not carry any relevant information for the 110 kV network are deleted, i.e. network elements with a voltage that is not 110 kV (e.g. EHV lines) or a frequency that is not 50 Hz (e.g. 16.7 Hz railway grid). The first stage of main-processing involves converting routes with multiple voltages and frequencies (e.g. routes carrying 380 kV and 110 kV systems) into unique voltage and unique frequency routes. Again, the connections are analysed to consider changes caused by deleted routes. Towers with implausible combinations of circuits are identified with the help of the pre-defined ruleset. All routes with implausible combinations are checked regarding their voltage and frequency information. The implausibility is attributable to routes with extra circuits which belong to non-110 kV voltage levels and non-50 Hz frequency. Therefore, the electrical information of such routes is modified iteratively for all towers to obtain a system consisting of routes with only 110 kV and 50 Hz.
The second stage of main-processing converts multi-circuit routes into single-circuit routes, since an electrical node-branch model has to consist of three phase circuits with three conductors. This is achieved by analysing the connection types and identifying all plausible combinations of circuits according to the pre-defined ruleset. Iteratively, towers with more than one circuit are processed until all multi-circuit routes are converted and a plausible node-branch model can be constructed. The π branch model is used to parameterize all branches of the developed network with characteristic series impedance and charging susceptance for sub-transmission lines and cables [8].
In the post-processing step the developed model consisting of 110 kV grid structures in Germany is integrated into a detailed model of the Continental European EHV grid and coupled via 380/110 kV and 220/110 kV transformers [9]. Again the parameters of the transformers are derived from typical electrical parameters based on their nominal power [10]. Finally, remaining electrical islands within the 110 kV system are eliminated by connecting them to the synchronous grid via synthetic lines.

2.3. Regionalization

The simulation of power flows requires detailed information about the generation capacities and loads at each grid node. The determination of these capacities based on overall installed capacities for each generation technology and demand is subject of the regionalization. The regionalization can be divided into two steps. The first one consists of the spatial allocation of generation and load capacities. The assignment of these capacities to grid nodes is subject of the second step. The allocation of renewable generation capacities, such as wind and photovoltaic, is done based on detailed meteorological and topological information for Europe [11]. Conventional power plants are allocated according to their location and connected voltage level [12]. The load is distributed based on detailed regional socio-economical information taking electric, heat and gas demand of private households, industry and commerce, trade & services (CTS) into account [13][14]. In Germany, the regionalization is done at postal code (ZIP) level. For this purpose two approaches are applied for the assignment of distributed generation units and loads to grid nodes (second step). The first approach assigns the information at ZIP level to each grid node proportional to the inverse of the distance between the center of the postal code region and the grid node. The number of nodes considered for the distribution has to be defined a priori. Here, the three nearest stations are considered for the distribution of capacities as shown in Figure 2, left, which also describes the share of capacities in a ZIP zone assigned to a station. This regionalization approach can be applied to both grid structures, with and without the 110 kV network.

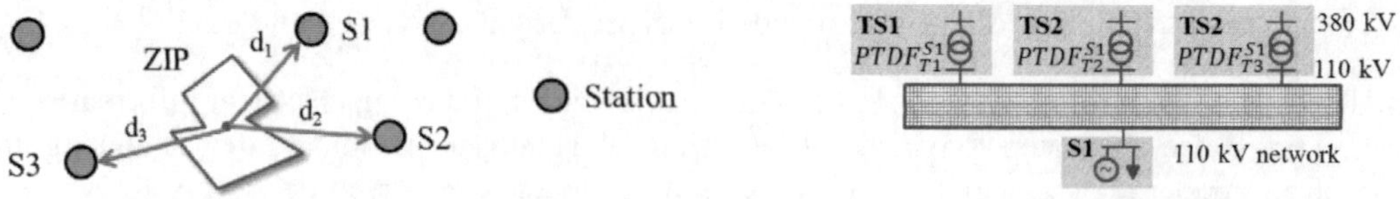

$$P_{S1,ZIP} = \frac{d_1^{-1}}{d_1^{-1} + d_2^{-1} + d_3^{-1}} * P_{ZIP} \qquad P_{TS1,S1} = \frac{PTDF_{T1}^{S1}}{\sum_t PTDF_t^{S1}} * P_{S1}$$

Figure 2. Assignment of capacities in ZIP zones to grid stations.

IOP Conf. Series: Journal of Physics: Conf. Series **977** (2018) 012004 doi:10.1088/1742-6596/977/1/012004

The second regionalization approach is applied for power flow simulations without consideration of the 110 kV network. However, the structure of the 110 kV network is used in order to derive installed capacities at each station of the transmission grid (220 kV or 380 kV). First, the regionalization is done considering all stations of the 110 kV network. Power Flow Distribution Factors (PTDF) are applied in order to distribute the regionalized capacities in subordinate grids to stations in the transmission grid. Figure 2, right, shows schematically the distribution of load and generation capacities connected to a 110 kV station, which is distributed among three stations in the transmission grid. $\mathrm{PTDF}_{\mathrm{T1}}^{\mathrm{S1}}$ defines the change in power flow on transformer T1 in station TS1 for a power injection at the 110 kV station S1. The share of capacity assigned to TS1 is done proportional to PTDF values of the network coupling transformers. This approach uses PTDF values as a metric to define the electrical proximity between the stations and allows a more accurate regionalization for the transmission network model (voltage level above 220 kV).

3. Model verification

The derived 110 kV grid model based on OSM data consists of 7,689 nodes (3,760 substations assumed to have coupled busbars and 3,529 connections outside substations, i.e. tripods) and 11,838 branches (OHL and cables) with a total circuit length of 77,156 km. This corresponds to 80.1% of the total circuit length according to official sources [15]. 148 of those branches are synthetic lines that connect 67 electrical islands consisting of 248 nodes. The generated network has been visually compared with published grid maps of German Distribution System Operators (DSO), e.g. Westfalen Weser Netz, LEW Verteilnetz, E.DIS, Stromnetz Berlin, Stromnetz Hamburg, inetz, TEN Thüringer Energienetze or Infrawest. The analysis exposed a close match in suburban and rural regions. Due to the high density of underground cables in urban regions and insufficient mapping in OSM, primarily the representation of those regions is more inaccurate and requires additional modelling in future work. Figure 3, left, shows the derived 110 kV system in Germany integrated in the European transmission grid [9]. Figure 3, right, visualizes the regionalization of generation according to subsection 2.3 and shows the installed capacities of renewable and conventional generation at each extra high voltage and high voltage grid node in Germany.

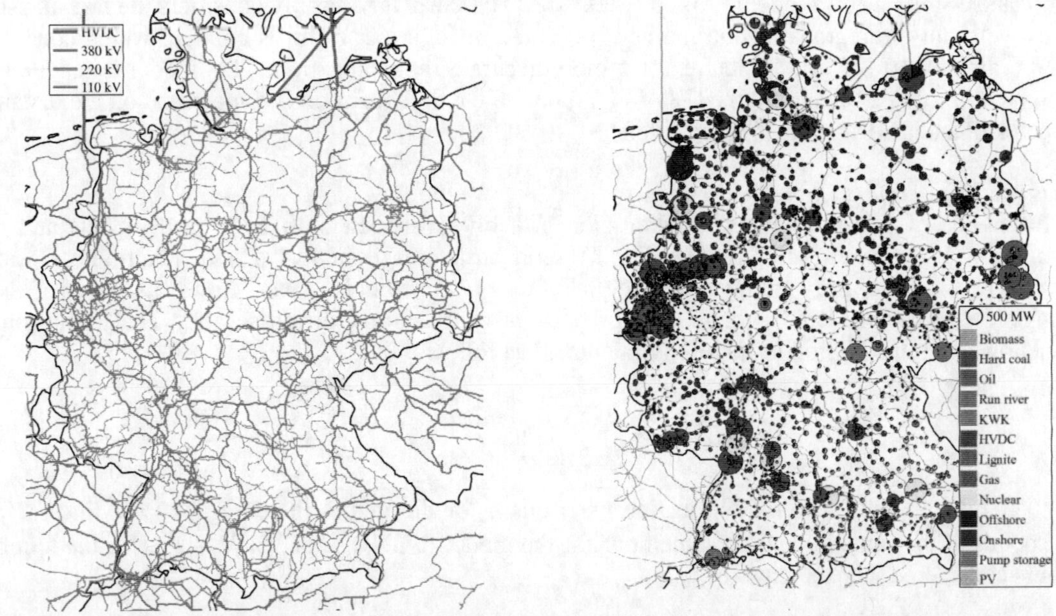

Figure 3: 110 kV grid model in Germany integrated into the European transmission system (left) and regionalization of conventional and renewable generation capacities (right).

4. Results

4.1. Approach and scenario framework

The impact of 110 kV grid structures on the congestion management in Germany is quantified by comparing different grid models and methods for regionalization of demand and renewable as well as conventional generation (cf. subsection 2.3). Demand time series are created taking standard electric load profiles and ENTSO-E data into account. Time series of renewable generators are generated based on the allocation of generation capacities (regionalization), meteorological data of the weather year 2012 and standard generation unit models [11]. To identify the power plant dispatch for 8760 hours, a simulation of the European electricity market is performed [16]. Using a Newton-Raphson based AC power flow algorithm, loadings of all transmission lines and transformers in the European transmission grid as well as in the German 110 kV grid are calculated. Subsequently, redispatch simulations are performed to quantify congestion management efforts.

The redispatch model is a linear optimization that minimizes total redispatch costs while dissolving all bottlenecks in the transmission grid including potential N-1 contingencies. The network topology is represented by linear sensitivities that indicate changes of power flows on branches in case of changes of injected power at nodes as well as in case of line outages. To keep the power system in balance, it has to be ensured that the increased and decreased power is equal when adjusting power of generation. The redispatch simulation includes conventional power plants, renewables and cross-border units. The parametrization is done according to the current regulatory framework in Germany (e.g. prioritized infeed from renewables) [17].

Since the 110 kV grid model generated by OSM data does not contain future grid expansion measures, a contemporary scenario for the year 2018 is chosen. The scenario framework regarding the installed generation capacities applied in this paper is based on officially published power plant lists, the Mid-term Adequacy Forecast by ENTSO-E and forecasts regarding the development of renewable energy sources in Germany [12][18][19].

The comparison of power flow as well as redispatch simulations is done on the basis of three grid models. The first transmission grid model includes the derived 110 kV grid structure, called here '110kV'. The second model only considers the transmission grid with line voltage levels equal to 220 kV and 380 kV. This model is called 'EHV_{DIST}'. The third variant only considers the transmission grid as well, but the regionalization method presented in Figure 2, right, is applied, which considers the subordinate 110 kV grid structure for a more accurate regionalization. This model is labelled as 'EHV_{PTDF}'. For the '110kV' and the 'EHV_{DIST}' model the regionalization is done based on the distance between ZIP and grid stations (Figure 2, left).

4.2. Power flow results

The difference of power flows on transmission grid lines between both EHV grid models and the transmission grid model including the 110 kV grid structures is analysed and quantified. Besides power flows in the pre-contingency state (Figure 4), maximal power flows in case of N-1 contingencies are considered (Figure 5). For both cases, the mean difference between the absolute power flows on transmission grid line l is calculated as follows:

$$\Delta S_{l,mean}^{EHV} = \frac{1}{H}\sum_{h=1}^{H}\left(\left|S_{l,h}^{110kV}\right| - \left|S_{l,h}^{EHV}\right|\right) \quad (1)$$

$S_{l,h}^{110kV}$ is the apparent power flow on line l in hour h for the model including 110 kV lines. $S_{l,h}^{EHV}$ corresponds to the same line in the same hour for the model 'EHV_{DIST}' or 'EHV_{PTDF}'. H is the number of hours considered, here equal to 8760.

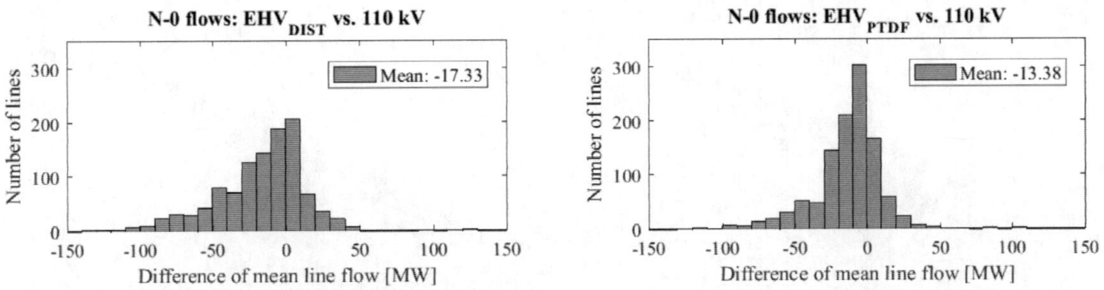

Figure 4. Distribution of the mean difference on power flows in the N-0 case for each line.

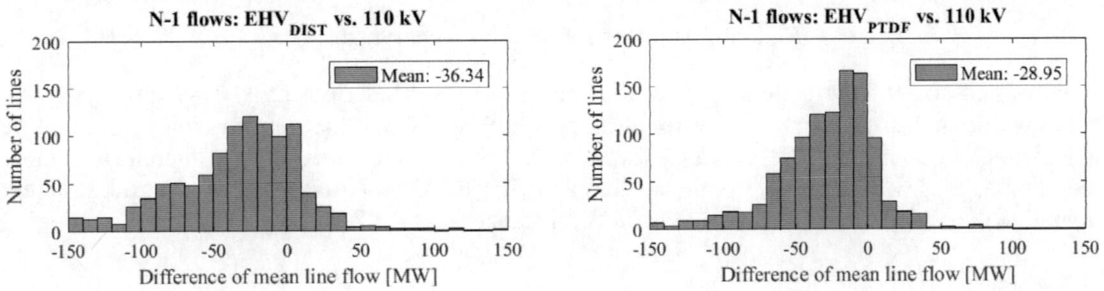

Figure 5. Distribution of the mean difference on power flows in the N-1 case for each line.

The results show an overall reduction of power flows in the transmission grid, most notably in the N-1 case, when considering the 110 kV grid structures. The 110 kV voltage level takes over a part of the power transfer and thus relieves the transmission grid. A visualization of this comparison is also shown in Figure 6. Positive values (green lines) represent a lower flow on the line due to the consideration of the 110 kV grid. This emphasizes the overall relieve of the transmission grid, but also proves increasing power flows in some specific cases.

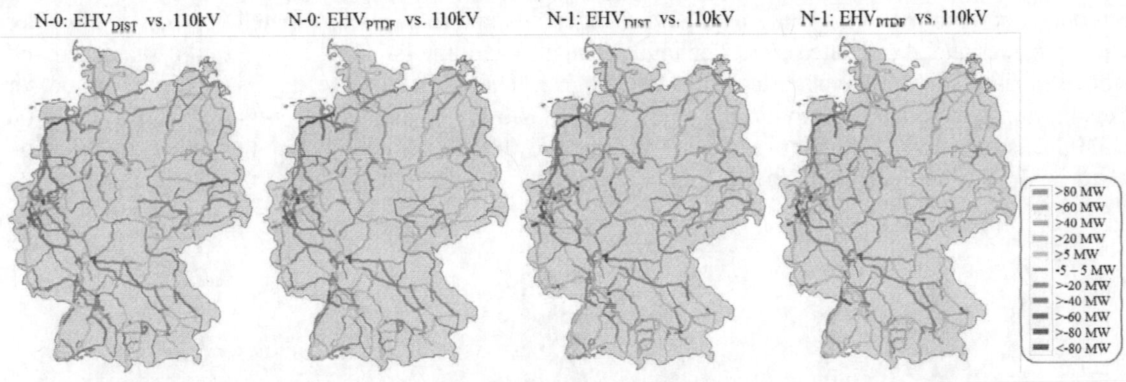

Figure 6. Line specific mean difference in power flows on the transmission grid lines.

Furthermore, the vertical load, i.e. the power exchange between the high and extra high voltage levels over the coupling transformers, is quantified. Figure 7 shows a comparison of the vertical load for each hour of the year. The vertical load is systematically higher when neglecting the 110 kV grid. This evidences that a consideration of the 110 kV grid structure leads to a lower energy flow between the transmission grid and the subordinate voltage levels. The loads at the 110 kV level can be supplied by generation located at the same voltage level such that power flows on transmission grid lines are partially superseded.

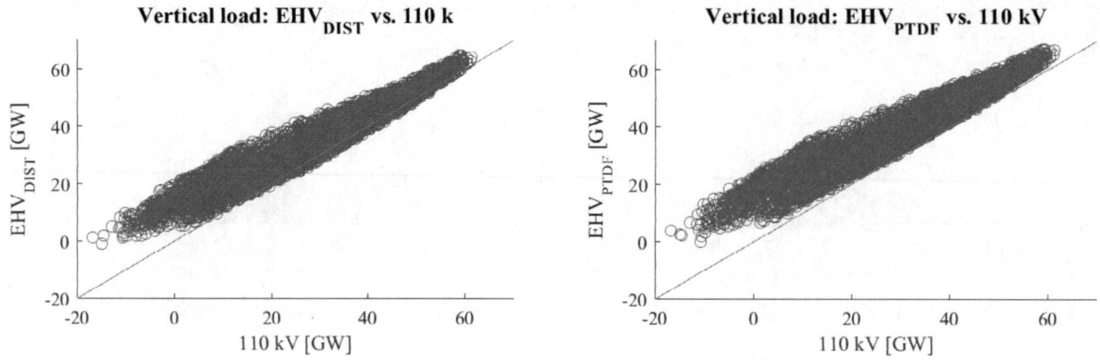

Figure 7. Comparison of the vertical load.

The analysis of power flow patterns proves the necessity for considering the 110 kV grid system in future network analysis. Otherwise, the resulting power flows within the transmission grid might be overestimated. Furthermore, the results show also a lower overestimation and therefore a more accurate modelling when applying regionalization for the EHV based on the 110 kV grid structure compared to the regionalization based solely on the distance between ZIP and grid nodes.

4.3. Redispatch results

Based on the market results and power flow calculations redispatch simulations are performed to quantify and compare the congestion management effort to relieve bottlenecks. Due to the large number of renewable units connected to the 110 kV network, wind onshore and photovoltaic units are aggregated to 300 groups per technology to reduce the complexity of the optimization when considering subordinate grid structures. A k-means clustering is applied to form these groups according to their node-branch sensitivity on all lines in the network.

Figure 8 shows the sum of power increases and decreases for all considered technologies for the three different grid models defined in section 4.2. The redispatch results confirm the results from chapter 4.2 and show a significant impact of 110 kV structures on the congestion management in Germany. Due to reduced power flows, especially for N-1 contingencies, the required redispatch volumes are smaller when considering subordinate grids. The total redispatch volume is 28.23 TWh (sum of increased and decreased power) when considering the 110 kV grid. The choice of the regionalization method for transmission grid models without the 110 kV level has a lower impact, but regionalization based on PTDF leads to a slightly smaller redispatch volume (43.08 TWh) in comparison to the regionalization based on the distance (45.61 TWh).

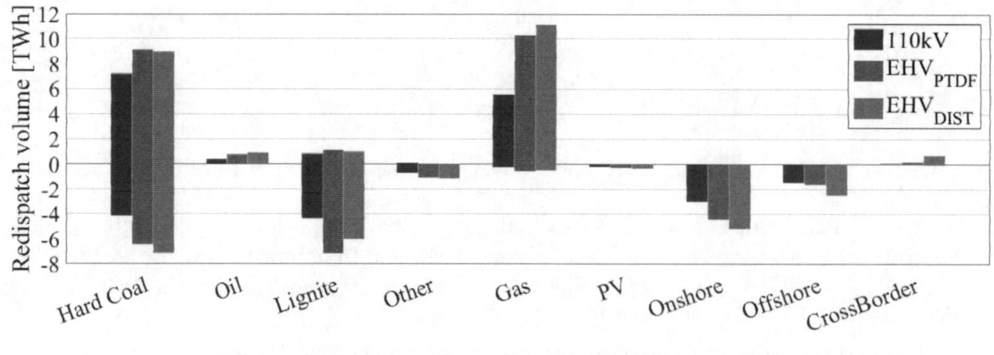

Figure 8. Redispatch results of analysed scenarios.

The regional distribution of the required redispatch is visualized in Figure 9. The comparison of all simulations shows similar redispatch patterns, but reduced power adjustments when considering subordinate grid structures. The 110 kV grid increases the transfer capacity within the system and allows the application of more detailed regionalization methods. This influences redispatch measures especially in weakly meshed regions dominated by 220 kV lines (e.g. in north-western part of Germany near Emden), which have notably lower thermal ratings than 380 kV lines.

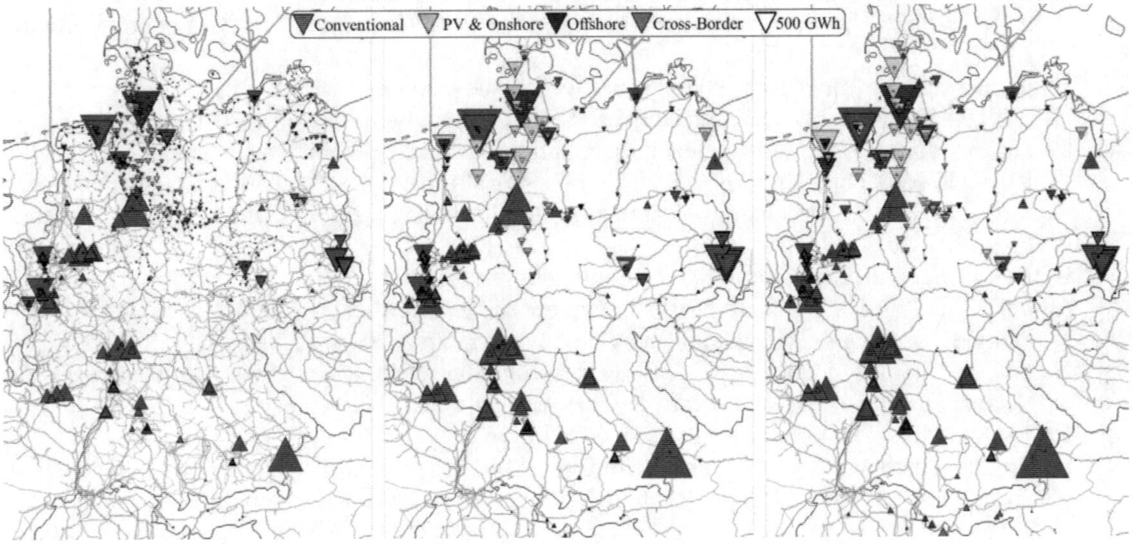

Figure 9. Geographical visualization of power increases (upward triangle) and decreases (downward triangle) for 110 kV (left), EHV_{PTDF} (middle), EHV_{DIST} (right).

5. Conclusion and outlook

The simulation of redispatch measures in Germany is significantly impacted by the modelling of 110 kV grid structures. Since power flows in the transmission grid, especially in case of N-1 contingencies, are systematically lower when considering 110 kV grid structures, fewer adjustments are required to relieve overloads. Since OSM includes partially insufficient data, future work will focus on further development of methods to process or complement grid information, e.g. the representation of cable-dominated urban regions. Additionally, the methods will be applied to derive models of subordinate grid structures of further European countries and quantify the impact on power flow as well as redispatch simulations for the whole European power system.

References

[1] Netztransparenz.de: EEG-Anlagenstammdaten 2015. Last check: 11/01/2016.

[2] 50Hertz Transmission GmbH; Amprion GmbH; TenneT TSO GmbH; TransnetBW GmbH: Netzentwicklungsplan Strom 2030, Version 2017, 1. Entwurf, 2017.

[3] Semerow, A. et. al.: Dynamic Study Model for the interconnected power system of Continental Europe in different simulation tools, IEEE Power Tech Proc., Eindhoven, 2015.

[4] Altschaffl, S., Witzmann, R., Ahndorf, T.: Generating a PSS™NETOMAC model of the German Transmission Grid from Google Earth and visualizing load flow results, IEE International Energy Conference (ENERGYCON), Dubrovnik, 2014.

[5] Villella, F., Leclerc S., Erlich, I., Rapoport, S.: PEGASE pan-European test-beds for testing of algorithms on very large scale power systems, ISGT Europe, Berlin, 2012.

[6] Boiarchuk, I., Hewes, D., Altschaffl, S., Witzmann, R.: Simulation Framework for Analysis of the European Transmission System under a Wide Range of Operating Conditions, 14.

Symposium Energieinnovation, Graz, 2016.
[7] Medjroubi, W. et al.: Open Data in Power Grid Modelling: New Approaches Towards Transparent Grid Models. Energy Reports,Volume 3, November 2017.
[8] Speck, D. et al.: Energiekabel im EVU, Renningen-Malmsheim, expert-Verlag, 1994.
[9] Natemeyer, H.: Scheufen, M.; Winter, S. et al.: Identification of environmentally relevant network hotspots. IEEE Power Tech Proc., Trondheim, 2011.
[10] ABB Calor Emag Schaltanlagen: Schaltanlagen, 10. Aufl., Berlin, 1999.
[11] Raths, S. et. al.: The Energy System Development Plan (ESDP), Int. ETG Congress, Bonn, 2015.
[12] Bundesnetzagentur, Bundeskartellamt: Kraftwerksliste. Last check: 12/20/2016.
[13] DDS: PLZ8 Deutschland XXL, Strukturdaten für kleinräumige Gebiete, Karlsruhe, 2011.
[14] Center for International Earth Science Information Network - CIESIN - Columbia University. 2016. Gridded Population of the World, Version 4 (GPWv4): Population Count. Palisades, NY: NASA Socioeconomic Data and Applications Center (SEDAC). Last check: 11/11/2016.
[15] Bundesnetzagentur, Bundeskartellamt: Monitoringbericht 2016, 2016.
[16] Müller, C. et. al.: Multimodale Marktsimulation zur ganzheitlichen Analyse des europäischen Energiesystems, 14. Symposium Energieinnovation, Graz, 2016.
[17] Natemeyer, H.: Modellierung der Betriebsführung elektrischer Übertragungsnetze für Netzplanungsprozesse, PhD thesis, RWTH Aachen, Aachen, 2017.
[18] ENTSO-E: Mid-term Adequacy Forecast, 2016 edition.
[19] Leipziger Institut für Energie GmbH: Mittelfristprognose zur deutschlandweiten Stromerzeugung aus EEG-geförderten Kraftwerken für die Kalenderjahre 2017 bis 2021, 2016.

IOP Conf. Series: Journal of Physics: Conf. Series **977** (2018) 012005 doi:10.1088/1742-6596/977/1/012005

Joint optimisation of arbitrage profits and battery life degradation for grid storage application of battery electric vehicles

Alexander Kies

Frankfurt Institute for Advanced Studies, Ruth-Moufang-Str. 1, 60439 Frankfurt am Main, Germany

E-mail: kies@fias.uni-frankfurt.de

Abstract. To meet European decarbonisation targets by 2050, the electrification of the transport sector is mandatory. Most electric vehicles rely on lithium-ion batteries, because they have a higher energy/power density and longer life span compared to other practical batteries such as zinc-carbon batteries. Electric vehicles can thus provide energy storage to support the system integration of generation from highly variable renewable sources, such as wind and photovoltaics (PV). However, charging/discharging causes batteries to degradate progressively with reduced capacity.

In this study, we investigate the impact of the joint optimisation of arbitrage revenue and battery degradation of electric vehicle batteries in a simplified setting, where historical prices allow for market participation of battery electric vehicle owners.

It is shown that the joint optimisation of both leads to stronger gains then the sum of both optimisation strategies and that including battery degradation into the model avoids state of charges close to the maximum at times. It can be concluded that degradation is an important aspect to consider in power system models, which incorporate any kind of lithium-ion battery storage.

1. Introduction

European power systems are transforming. Changes towards renewable generation are driven by aims of decarbonisation and sustainability. However, wind and solar power have intermittent generation profiles, which makes their power system integration challenging [1, 2]. Various options have been proposed in the past as contributions to the balance of generation and demand in the system: Optimising the mix of generation from different renewable power sources such as wind, solar or hydro power [3, 4, 5, 6, 7, 8], storage [9, 10, 11], transmission grid extensions [12, 13], or concentrated solar power imports from North Africa, which are attractive because of the ability to store energy in form of heat [14]. Other options include system-friendly renewables or load shifting [15, 16, 17, 18, 19].

As electric cars are getting more and more adequate for daily use due to significant cost reduction and driving range extension, the deployment of electric vehicles in the EU is gaining momentum [20]. Automotive companies like Daimler expect up to one in four cars sold to be electric by 2025 [21]. This process comes with a major opportunity: Electric vehicles are equipped with batteries, which therefore hold aggregated large amounts of energy capacity storage. Two

IOP Conf. Series: Journal of Physics: Conf. Series **977** (2018) 012005 doi:10.1088/1742-6596/977/1/012005

$b(t)$	charging rate of battery
$\hat{b(t)}$	charging capacity of battery
$\tilde{b(t)}$	charging due to driving
λ_L	price for degradation, i.e., relative share of investment cost
$\lambda(t)$	time series of prices
$L(t)$	life status of battery
ΔL	battery degradation relative to new battery
L_{cal}	life status or aging due to calendaric processes
L_{cyc}	life status or aging due to cycling processes
l_c	aging of a single cycle
σ	state of charge
σ_0	reference state of charge
σ^+	battery energy capacity
δ	depth of discharge
S	stress factor
k	constant of stress factor model
t	time index
$\bar{T}$	average temperature

Table 1: Important terminology

different options are exploitable for electric vehicles to balance power systems [22]: grid to vehicle (G2V) operation, i.e., adapting charging patterns to the generation side and vehicle to grid (V2G) operation, i.e., feeding electricity back into the grid. This sector coupling (usually, the heating sector is considered as well) is often expected to be a key ingredient to a decarbonised energy system. The coupling of the sectors electricity, heat and transport was investigated on the European scale in [23, 24] and found to provide significant cost reduction potential.
In this paper, the interaction of the battery electric vehicle and the electricity grid is investigated. It is assumed that a battery is discharged due to driving (based on an averaged demand profile) and can charge/discharge at hourly prices. The comparison of different charging strategies, e.g., with and without taking degradation into account, allows to study the relevance of degradation for batteries participating in energy markets. Furthermore, results indicate that the benefit of jointly optimising arbitrage and degradation leads to stronger benefits than the sum of both independently. The nomenclature used throughout this paper is shown in Table 1.

2. Methodology

To calculate discharging of batteries due to regular operation, i.e., driving and parking, we use traffic data that is provided by the Bundesanstalt für Straßenwesen (BASt) [25] and consists of hourly measurements for the year 2015 of approximately 1700 stations distributed along the German road network. First, this data is further simplified to obtain an averaged weekly demand profile. This weekly demand profiles as used in the following calculations can be seen in Fig. 1. In principle, it would be possible to take the data directly without averaging it further. However, to increase comparability with other works, which often rely on weekly profiles (e.g., [26]), we have adopted our data in an appropriate way. For prices, we use three years of historical EEX price data (2012-2015) [27]. This dataset consists of hourly day-ahead spot prices. A battery owner is assumed to be able to trade at hour t at corresponding price $\lambda(t)$.

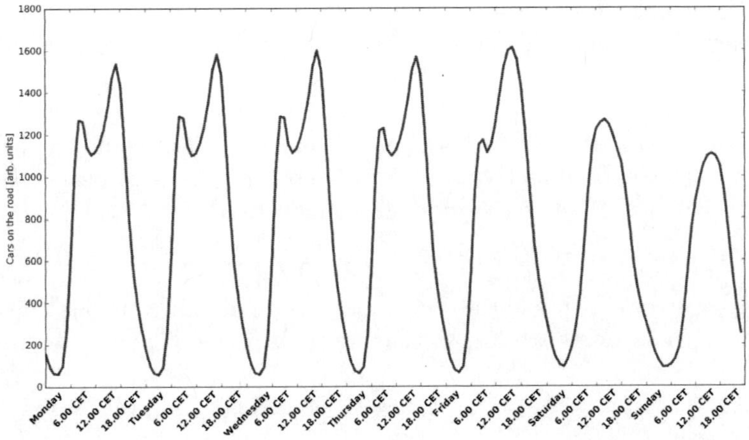

Figure 1: Weekly traffic profile for Germany derived from measurements by BASt.

3. Battery life degradation

Lithium-ion battery aging is a complex process [28] that is not fully understood on the microscopic level. Processes involved include, for instance, solid-electrolyte interphase film formation [29, 30]. Caused by the lack of understanding degradation processes, battery lifetime predictions are usually based on empirical models rather than bottom-up mathematical modelling.

The degradation model used in this work is adapted from the work by Millner [31] and was further refined by Xu et al. [32].

The state of a battery is characterised by the parameter L, where $L = 0$ refers to a new battery and $\Delta L = 1 - L(t)$ describes the remaining energy capacity. Battery aging in a temporal interval from t_0 to t_1 is assumed to consist of cycle and calendar aging,

$$L([t_0, t_1]) = L_{\text{cal}}([t_0, t_1]) + L_{\text{cyc}}([t_0, t_1]). \tag{1}$$

Calendar aging is a function of the state of charge (SOC) σ, $L_{\text{cal}} = L_{\text{cal}}([t_0, t_1], \sigma, \bar{T})$, and cycle aging the sum over aging caused by the n single cycles within the interval,

$$L_{\text{cyc}} = \sum_{i=1}^{n} n_i l_c(\sigma_i, \delta_i, \bar{T}_i). \tag{2}$$

$n_i \in \{0.5, 1.0\}$ indicates half (0.5) or full cycles (1.0). It is furthermore assumed, that aging processes can be expressed via a product of different stress factors. In the case of cycle aging this reads

$$l_c(\delta, \sigma, \bar{T}) = S_\delta(\delta) S_\sigma(\sigma) S_T(\bar{T}), \tag{3}$$

and in the case of calendar aging

$$L_{\text{cal}}([t_0, t_1], \sigma, \bar{T}) = S_t([t_0, t_1]) S_\sigma(\sigma) S_T(\bar{T}). \tag{4}$$

The stress factors of aging due to temperature, S_T, and of aging due to passing of time, S_t, are not considered here in an explicit way, because we focus on investigating the main behaviour

of different optimisation strategies. However, if realistic values of degradation ought to be computed, these need to be taken into account, as well.
For the state of charge stress model, we adapt the exponential approach from Millner ([31]),

$$S_\sigma = e^{k_1(\sigma - \sigma_0)}, \tag{5}$$

where σ_0 is a reference state of chart, chosen in this work to be 0.5 and k_1 a constant factor, which can be determined experimentally. For the depth of discharge stress model, different approaches can be found in literature. Besides the quadratic one,

$$S_\delta(\delta) = k_2\delta^2 \tag{6}$$

that is adapted in this work, exponential and modified quadratic versions exist [32, 33]. The best fitting model depends on the battery chemistry. However, they all exhibit very similar behaviour [32].
To measure irregular charging cycles, the rainflow-cycle counting algorithm [34, 35] as proposed by Xu et al. [32] is used.
It can be summarised for a time series as follows:

- 1. identify local maxima and minima
- 2. find global maximum and minimum (count as half-cycle)
- 3. wlog: if global maximum first (in the case of global minimum first, act vice versa):
  - 1. count half cycles from the global maximum to the most negative minimum before it, then the most positive maximum, etc.
  - 2. count half cycles after the global minimum until the most positive maximum, etc.
  - 3. identify all remaining as full cycles

Other degradation models are often simplified to linear or quadratic relationships [26, 36, 37]. This is strongly motivated by the fact that power system optimisation is often performed using linear models.

4. Results

The potential benefits of different optimisation schemes with different objectives are investigated. The benchmark is regular operation, i.e., cars are charged once they are below a certain threshold, which is chosen to be 0.8. 0.8 is a good tradeoff between fully charging the battery and hence exposing it to significant SOC stress and keeping at the same time a relatively high amount of energy at storage. If their charging level without grid interaction, $\tilde{\sigma}(t) = \sigma(t-1) + b_{\text{discharge}}(t)$, where $b_{\text{discharge}}$ is the discharge due to driving, falls below this level, the battery is charged,

$$b(t) = \min\{\hat{b}, 0.8 - \tilde{\sigma}(t)\}, \tag{7}$$

where $\hat{b}$ is the charging capacity. If $b(t) - b_{\text{discharge}}(t) < 0$, the battery feeds electricity into the grid. It is assumed that cars can charge at all times. However, since we consider an average car battery (e.g., via averaging demand of cars), this seems an appropriate simplification. The overall cost is composed from the cost of charging and battery degradation,

$$\text{Cost} = \sum_t b(t) \cdot \lambda(t) + \Delta L \cdot \lambda_L. \tag{8}$$

The first strategy aims at maximising arbitrage revenue via the injection $b(t)$ and reads

$$\begin{aligned} \underset{b(t), t\in[t_0, t_0+72h]}{\text{minimise}} \quad & \lambda(t)b(t) \\ \text{subject to} \quad & 0 \le \sigma(t) \le \sigma^+ \\ & |b(t)| \le 0.15\sigma^+, \end{aligned} \tag{9}$$

where $\lambda(t)$ is the price time series taken from historical spot price data for Germany. The charging strategy is always optimised under the assumption of perfect foresight for 72 hours, consecutively for the next 72 hours, etc. Furthermore, at the beginning/end of every cycle, state of charge is constrained to 0.5. In addition, we assume the reference state of charge to be 0.5 (as recommended by Xu et al. [32]) and put limits on the charging/discharging of 15% of the energy capacity. The second strategy simply aims at reducing overall battery degradation. It reads

$$\begin{aligned} \underset{b(t),t\in[t_0,t_0+72h]}{\text{minimise}} \quad & \lambda_L \Delta L \\ \text{subject to} \quad & 0 \leq \sigma(t) \leq \sigma^+ \\ & |b(t)| \leq 0.15\sigma^+. \end{aligned} \tag{10}$$

The final strategy combines both strategies by jointly maximising arbitrage revenue and minimising battery degradation:

$$\begin{aligned} \underset{b(t),t\in[t_0,t_0+72h]}{\text{minimise}} \quad & \left[\sum_t \lambda(t)b(t) + \lambda_L \Delta L\right] \\ \text{subject to} \quad & 0 \leq \sigma(t) \leq \sigma^+ \\ & |b(t)| \leq 0.15\sigma^+. \end{aligned} \tag{11}$$

We assume that all batteries charge/discharge following the same patterns. This can be shown to be the Nash-Equilibrium.
Figure 2 shows state of charge (a), charging rates (b) as well as the corresponding time series of prices (c) for two optimisation strategies in an arbitrary 72-hour period. Discharging due to regular operation was removed from the charging rates. In general, three price peaks and valleys can be observed with the first two being especially prominent with prices reaching values above 50 Euros / MWh, thus leading to peak-valley differences of 30 Euros / MWh and more. The last peak is less pronounced with a peak-valley difference of just around 10 Euro / MWh. The state of charge is shown for charging strategies three (maximising arbitrage revenue) and four (maximising arbitrage revenue under consideration of degradation) for the same period. It can be seen that taking degradation into account has the largest effect on the charging behaviour during the third peak-valley period, where price spreads are comparably small. In this case, not the full battery capacity is exploited. Instead, state of charge remains below the minimum, therefore reducing stress factors due to state of charge and depth of discharge.
The cost results averaged over the entire three years are depicted in Fig. 3. The case without optimisation works as the benchmark. If solely degradation is minimised, a cost reduction of 8% can be observed. This is reasonable, although it is a simple task to construct a price time series which would lead to a cost increase, if degradation was minimised without considering prices. Maximising arbitrage revenue without considering battery degradation leads to a cost reduction of approximately 19%. If both are jointly optimised, the resulting cost reduction equals approximately 38%. Therefore, the reduction by joint optimisation outperforms the sum of cost reductions by optimisation of the single terms. A similar result was found by Shi et al. in the case of peak shaving and arbitrage revenue maximisation and coined “super linear gain”[38].

5. Critical appraisal

Several simple assumptions were made within this work, which could be replaced in future work to yield results closer to reality. In the future, price signal statistics are expected to change due to quickly rising shares of renewables and their intermittent feed-in profiles as well as increased flexibility of the demand side due to price signals reaching consumers. Hence, it seems reasonable

IOP Conf. Series: Journal of Physics: Conf. Series **977** (2018) 012005 doi:10.1088/1742-6596/977/1/012005

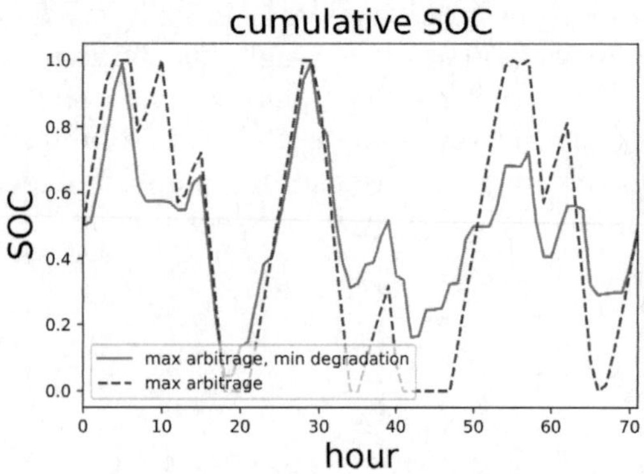

(a) State of charge (SOC) under joint optimisation of arbitrage revenue and degradation (red) and optimisation of arbitrage revenue alone (blue).

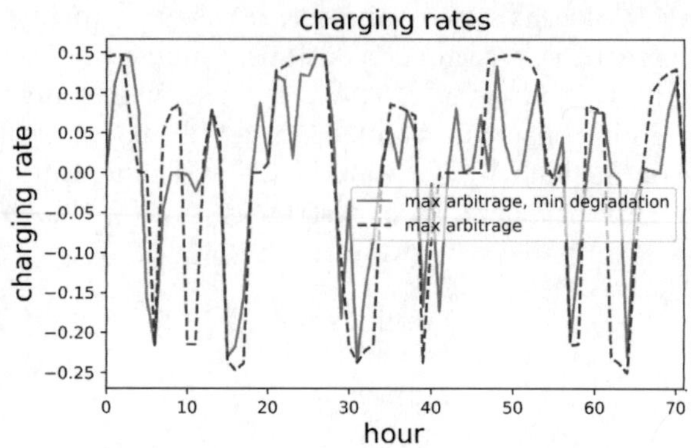

(b) Charging of battery under joint optimisation of arbitrage revenue and degradation (red) and optimisation of arbitrage revenue alone (blue).

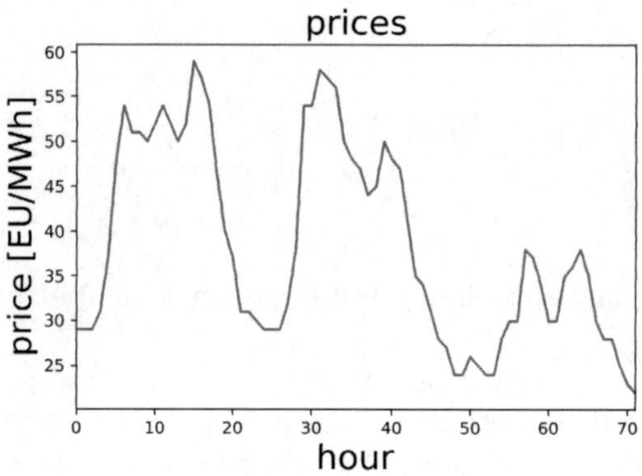

(c) Price time series.

Figure 2: All time series show the same exemplary time period of 72 consecutive hours in hourly resolution. Discharging due to driving was subtracted from the charging rate in Fig. b).

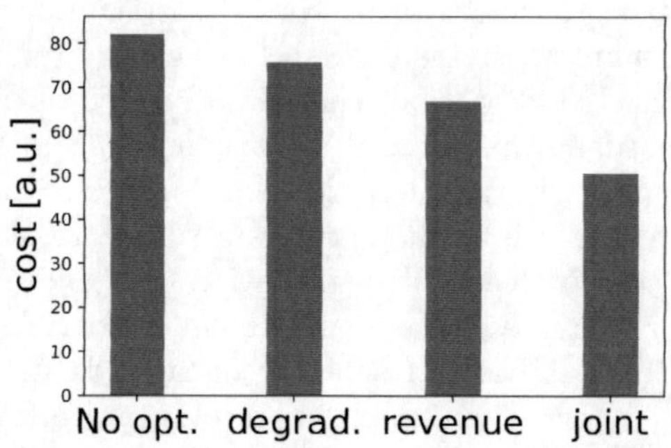

Figure 3: Cost of charging and degradation minus arbitrage revenue from market participation for the different charging strategies.

to study the potential gains of market participation of electric vehicle battery owners in a highly renewable power system, where prices can be obtained using shadow prices in an investment optimisation program.

Furthermore, the battery chemistry was not specified further, besides the assumption of the quadratic model for depth of discharge degradation. Specifying the battery chemistry could allow for realistic numbers on prices and degradation constants and consequently statements on future real-live operation and its cost performance.

Several heuristic assumption on the charging level were made. These include, for example, that the SOC at the beginning and end of each period is constrained to 0.5. Investigating the sensitivity of the results to the choice of this and other heuristic parameter is planned as a future extension of this work.

6. Summary, conclusion and outlook

In this work, the joint optimisation of arbitrage revenue and battery degradation was investigated in a simplified setting. It was shown that it leads to higher gain than the sum of the optimisation of the two processes separately. The minimisation of degradation alone without consideration of variable prices leads to a reduction of approximately 8% in overall cost. In principle, one could easily construct a case where this strategy would increase overall cost. However, it seems that prices are sufficiently strong correlated to discharging processes and thus allow for a reduction by degradation minimisation alone.

Semi-empirical lithium-ion battery degradation models like the one used in this work show that degradation is smallest when the state of charge is close to average and deep discharges are avoided. These non-linearities make the integration into power system optimisation models, which are for obvious reasons commonly formulated in a linear way, difficult. However, since investment into battery electric vehicles happens independently of grid storage application, it seems reasonable to model their operation with endogenous prices as was done in this work.

Another interesting question this research raises is that of the optimum sizing of batteries in electric vehicles. Typical combustion-based cars have gas tanks which might be equivalent to batteries with capacities of around 150 KWh. However, because degradation is reduced mostly close to the reference state of charge and deep discharge cycles should be avoided, it seems reasonable to expect optimum battery sizes to be larger than those values.

Finally, the underlying research concludes that the electrification of the transport sector contributes in two ways to the success of the European energy transition: i) by decarbonisation of the transport sector, ii) by allocation of storage to integrate variable renewable generation.

Acknowledgments

I thank my colleagues from FIAS, especially Tom Brown, David Schlachtberger and Jonas Hoersch for helpful suggestions. Furthermore, I thank Bruno Schyska from the University of Oldenburg as well. Additional gratitudes go to two anonymous reviewers for helpful comments and constructive criticism. A. Kies is financially supported by Stiftung Polytechnische Gesellschaft Frankfurt am Main.

References

[1] Henrik Lund. Large-scale integration of optimal combinations of pv, wind and wave power into the electricity supply. *Renewable Energy*, 31(4):503–515, 2006.

[2] Alexander Kies, Bruno U Schyska, and Lueder von Bremen. Curtailment in a highly renewable power system and its effect on capacity factors. *Energies*, 9(7):510, 2016.

[3] Alexander Kies, Bruno U Schyska, and Lueder von Bremen. The optimal share of wave power in a highly renewable power system on the iberian peninsula. *Energy Reports*, 2:221–228, 2016.

[4] Dominik Heide, Lueder Von Bremen, Martin Greiner, Clemens Hoffmann, Markus Speckmann, and Stefan Bofinger. Seasonal optimal mix of wind and solar power in a future, highly renewable europe. *Renewable Energy*, 35(11):2483–2489, 2010.

[5] B François, B Hingray, D Raynaud, M Borga, and JD Creutin. Increasing climate-related-energy penetration by integrating run-of-the river hydropower to wind/solar mix. *Renewable Energy*, 87:686–696, 2016.

[6] Jakub Jurasz and Bartłomiej Ciapała. Integrating photovoltaics into energy systems by using a run-off-river power plant with pondage to smooth energy exchange with the power gird. *Applied Energy*, 198:21–35, 2017.

[7] Alexander Kies, K Nag, L von Bremen, Elke Lorenz, and Detlev Heinemann. Investigation of balancing effects in long term renewable energy feed-in with respect to the transmission grid. *Advances in Science and Research*, 12(1):91–95, 2015.

[8] Jakub Jurasz. Modeling and forecasting energy flow between national power grid and a solar–wind–pumped-hydroelectricity (pv–wt–psh) energy source. *Energy Conversion and Management*, 136:382–394, 2017.

[9] Guido Pleßmann, Matthias Erdmann, Markus Hlusiak, and Christian Breyer. Global energy storage demand for a 100% renewable electricity supply. *Energy Procedia*, 46:22–31, 2014.

[10] Dominik Heide, Martin Greiner, Lueder Von Bremen, and Clemens Hoffmann. Reduced storage and balancing needs in a fully renewable european power system with excess wind and solar power generation. *Renewable Energy*, 36(9):2515–2523, 2011.

[11] Stefan Weitemeyer, David Kleinhans, Lukas Wienholt, Thomas Vogt, and Carsten Agert. A european perspective: potential of grid and storage for balancing renewable power systems. *Energy Technology*, 4(1):114–122, 2016.

[12] Florian Steinke, Philipp Wolfrum, and Clemens Hoffmann. Grid vs. storage in a 100% renewable europe. *Renewable Energy*, (50):826–832, 2013.

[13] Tom Brown. Transmission network loading in europe with high shares of renewables. *IET Renewable Power Generation*, 9(1):57–65, 2014.

[14] Franz Trieb, Christoph Schillings, Thomas Pregger, and Marlene O'Sullivan. Solar electricity imports from the middle east and north africa to europe. *Energy Policy*, 42:341–353, 2012.

[15] Lion Hirth and Simon Müller. System-friendly wind power: How advanced wind turbine design can increase the economic value of electricity generated through wind power. *Energy Economics*, 56:51–63, 2016.

[16] Kabitri Chattopadhyay, Alexander Kies, Elke Lorenz, Lüder von Bremen, and Detlev Heinemann. The impact of different pv module configurations on storage and additional balancing needs for a fully renewable european power system. *Renewable Energy*, 113:176–189, (2017).

[17] Omar Hafez and Kankar Bhattacharya. Optimal planning and design of a renewable energy based supply system for microgrids. *Renewable Energy*, 45:7–15, 2012.

[18] Alexander Kies, Bruno U Schyska, and Lueder von Bremen. The demand side management potential to balance a highly renewable european power system. *Energies*, 9(11):955, 2016.

[19] Peter Palensky and Dietmar Dietrich. Demand side management: Demand response, intelligent energy systems, and smart loads. *IEEE transactions on industrial informatics*, 7(3):381–388, 2011.

[20] C Thiel, J Krause, and P Dilara. Electric vehicles in the eu from 2010-2014-is full scale commercialisation near. *JRC Science and Policy Report*, 2015.

[21] Udo Hartmann. Sustainability management environment@daimler, 2016.

[22] Henrik Lund and Willett Kempton. Integration of renewable energy into the transport and electricity sectors through v2g. *Energy policy*, 36(9):3578–3587, 2008.

[23] Tom Brown, David Schlachtberger, Alexander Kies, and Martin Greiner. Sector coupling in a highly renewable european energy system. In *15th Wind Integration Workshop*, 2017.

[24] Katrin Schaber, Florian Steinke, and Thomas Hamacher. Managing temporary oversupply from renewables efficiently: Electricity storage versus energy sector coupling in germany. In *International Energy Workshop, Paris*, 2013.

[25] Automatische verkehrszaehlung 2015. http://www.bast.de. Accessed: 2016-07-30.

[26] Alexander Schuller, Benjamin Dietz, Christoph M Flath, and Christof Weinhardt. Charging strategies for battery electric vehicles: Economic benchmark and v2g potential. *IEEE Transactions on Power Systems*, 29(5), 2014.

[27] Eex spot prices. http://www.eex.com. Accessed: 2016-06-15.

[28] Thomas Waldmann, Marcel Wilka, Michael Kasper, Meike Fleischhammer, and Margret Wohlfahrt-Mehrens. Temperature dependent ageing mechanisms in lithium-ion batteries–a post-mortem study. *Journal of Power Sources*, 262:129–135, 2014.

[29] Matthew B Pinson and Martin Z Bazant. Theory of sei formation in rechargeable batteries: capacity fade, accelerated aging and lifetime prediction. *Journal of the Electrochemical Society*, 160(2):A243–A250, 2013.

[30] H Buqa, A Würsig, J Vetter, ME Spahr, F Krumeich, and P Novák. Sei film formation on highly crystalline graphitic materials in lithium-ion batteries. *Journal of power sources*, 153(2):385–390, 2006.

[31] Alan Millner. Modeling lithium ion battery degradation in electric vehicles. In *Innovative Technologies for an Efficient and Reliable Electricity Supply (CITRES), 2010 IEEE Conference on*, pages 349–356. IEEE,

IOP Conf. Series: Journal of Physics: Conf. Series **977** (2018) 012005 doi:10.1088/1742-6596/977/1/012005

2010.

[32] Bolun Xu, Alexandre Oudalov, Andreas Ulbig, Goran Andersson, and Daniel Kirschen. Modeling of lithium-ion battery degradation for cell life assessment. *IEEE Transactions on Smart Grid*, 2016.

[33] Guannan He, Qixin Chen, Chongqing Kang, Pierre Pinson, and Qing Xia. Optimal bidding strategy of battery storage in power markets considering performance-based regulation and battery cycle life. *IEEE Transactions on Smart Grid*, 7(5):2359–2367, 2016.

[34] M Matsuishi and T Endo. Fatigue of metals subjected to varying stress. *Japan Society of Mechanical Engineers, Fukuoka, Japan*, 68(2):37–40, 1968.

[35] Stephen D Downing and DF Socie. Simple rainflow counting algorithms. *International journal of fatigue*, 4(1):31–40, 1982.

[36] Miguel A Ortega-Vazquez. Optimal scheduling of electric vehicle charging and vehicle-to-grid services at household level including battery degradation and price uncertainty. *IET Generation, Transmission & Distribution*, 8(6):1007–1016, 2014.

[37] Michael Koller, Theodor Borsche, Andreas Ulbig, and Goran Andersson. Defining a degradation cost function for optimal control of a battery energy storage system. In *PowerTech (POWERTECH), 2013 IEEE Grenoble*, pages 1–6. IEEE, 2013.

[38] Yuanyuan Shi, Bolun Xu, Di Wang, and Baosen Zhang. Using battery storage for peak shaving and frequency regulation: Joint optimization for superlinear gains. *arXiv preprint arXiv:1702.08065*, 2017.

Extension algorithm for generic low-voltage networks

S Marwitz[1], **C Olk**[1]

[1] Fraunhofer Institute for Systems and Innovation Research (ISI), Breslauer Strasse 48, 76139 Karlsruhe, Germany

E-mail: `simon.marwitz@isi.fraunhofer.de`

Abstract.
Distributed energy resources (DERs) are increasingly penetrating the energy system which is driven by climate and sustainability goals. These technologies are mostly connected to low-voltage electrical networks and change the demand and supply situation in these networks. This can cause critical network states. Network topologies vary significantly and depend on several conditions including geography, historical development, network design or number of network connections. In the past, only some of these aspects were taken into account when estimating the network investment needs for Germany on the low-voltage level. Typically, fixed network topologies are examined or a Monte Carlo approach is used to quantify the investment needs at this voltage level. Recent research has revealed that DERs differ substantially between rural, suburban and urban regions. The low-voltage network topologies have different design concepts in these regions, so that different network topologies have to be considered when assessing the need for network extensions and investments due to DERs. An extension algorithm is needed to calculate network extensions and investment needs for the different typologies of generic low-voltage networks. We therefore present a new algorithm, which is capable of calculating the extension for generic low-voltage networks of any given topology based on voltage range deviations and thermal overloads. The algorithm requires information about line and cable lengths, their topology and the network state only. We test the algorithm on a radial, a loop, and a heavily meshed network. Here we show that the algorithm functions for electrical networks with these topologies. We found that the algorithm is able to extend different networks efficiently by placing cables between network nodes. The main value of the algorithm is that it does not require any information about routes for additional cables or positions for additional substations when it comes to estimating network extension needs.

1. Background and objective

Due to climate and sustainability goals, distributed energy resources (DERs) like photovoltaic (PV)-systems are increasingly contributing to the electricity generation in Germany. One goal for German society is to install 2.5 GW additional PV-power every year [7]. Furthermore, plug-in electric vehicles (PEVs) are an option to reduce green house gases, if these vehicles are charged by renewable energy sources. Therefore, the federal government implemented a buyer's premium for PEVs [3].

Economic incentives for PEVs and PV-systems could lead to an increasing market penetration for these technologies in Germany in the future [8, 22]. These technologies are mostly connected to low-voltage electrical networks and their operation can lead to high loads there. While load peaks can cause critical network states, distribution system operators (DSOs) are responsible for a stable and reliable network operation [15, 17, 19, 20]. In the long term DSOs respond to critical network states by investing into network assets. This is done by replacing or adding new cables, overhead lines or voltage transformation substations to existing networks. Due to the long life time cycles of these network assets certain planning aspects are being taken into consideration. The planning of electrical networks has a long history [11, 14, 16]. In general all planning methods focus on certain aspects like the substation locations and capacities, or the feeder system switching and routing. But not all aspects of network planning can be taken into account [21].

DSOs still require tools for better investment decisions under uncertainty, which also contribute to an affordable, sustainable and reliable electric energy supply [12]. Therefore, optimal distributed generation placement algorithms have been developed to avoid critical network states in the future [13, 23]. In order to find optimal solutions for network planning methods like mixed integer programming or ant colony algorithms are used [1, 9, 10]. In Germany, more than 800 DSOs are responsible for the stable and secure operation of roughly 1 million kilometres of low-voltage networks [4, 5]. Network topologies vary significantly and depend on several conditions including geography, historical development, network design or the number of network connections. Due to a large number of different networks where the local conditions are also often unknown, not every single network can be planned by these algorithms to estimate the network investment needs on a low voltage level for the following years. Therefore, in the past, when it comes to estimating these investment needs, fixed topologies or a Monte Carlo approach is used [2, 6]. For example, if voltage range deviations occur networks are extended by an additional cable from the transformer to the last third of the overloaded line. Or a new cable is added between a transformer and the middle of the overloaded line if a thermal overload occurs [2]. In order to get more precise investment estimations for the future, this method should be formalize for arbitrary extension points at generic network topologies. Therefore, we introduce in section 2 a new extension algorithm which is capable of extending existing generic networks based on thermal overloads and voltage range deviations at these networks. In section 3 we set up a case study to test the algorithm on networks of radial, loop, or meshed topology. Finally in section 4, we draw conclusions for the algorithm and the case study.

2. Extension algorithm based on voltage range deviations and thermal overloads

2.1. Algorithm overview

The algorithm presented in this paper adds cables to the parts of an existing network topology which faces the strongest overload. The length of the added cables and consequently the network extension can be set by adjusting a relief factor F_{relief} with $0 > F_{relief} \geq 1$. The algorithm checks the currents in all cables of the network $\mathcal{C}$ and the voltages at all nodes $\mathcal{N}$ of the network. Similar to [2] we differentiate between voltage range deviations and thermal overloads. Thermal overloads are a result of currents exceeding rated values in either cables or transformers. Voltage range deviations occur if the voltage at a given node crosses a threshold. For both situations, the algorithm attempts to bring the values back to the nominal value by adding a new cable to the network topology. The algorithm is designed to use the shortest path between the connections of a newly added cable. After a network extension, the network state is tested again for overloads and voltage range deviations. If an overload occurs new cables are added to the network until a network state without overloads is reached.

The algorithm is available at *GitHub*[1] and implemented within the programming language *Java*. Nevertheless, the algorithm works independent of the programming language.

[1] `https://github.com/ChrisOlk/GenericLVExtensionAlgorithm`

IOP Conf. Series: Journal of Physics: Conf. Series **977** (2018) 012006 doi:10.1088/1742-6596/977/1/012006

2.2. Network extension due to thermal overload

If a thermal overload occurs, the algorithm adds a new cable between two nodes in the network parallel to overloaded components. The algorithm finds that sequence of cables $\mathcal{S}_c$ which faces the thermal overloaded. To determine the extent of the overload of a cable, the actual current $I(c_i)$ is compared to the rated current of the cable $I_{rated}(c_i)$. In a second step, two nodes in $\mathcal{S}_c$ are found based on F_{relief}. Finally, at these nodes the network is extended.

2.2.1. Find the cable sequence with the highest thermal overload A search algorithm is deployed to find the serial sequence of cables $\mathcal{S}_c$ with the strongest thermal overload. Cables are added to $\mathcal{S}_c$ on the basis on their thermal overload $\left(\frac{I(c_i)}{I_{rated}(c_i)}\right)$ and the voltage at their connected nodes $V(n)$. First the cable $c_{I,max}$ is selected which holds the highest relative thermal overload $\left(\frac{I(c_i)}{I_{rated}(c_i)}\right)$. The search for other cables starts from that selected cable c_s.

$$c_s = c_{I,max} = \underset{c_i \in \mathcal{C}}{\arg\max} \left(\frac{I(c_i)}{I_{rated}(c_i)}\right) \tag{1}$$

Due to Ohms law one node $n_{c_s,high}$ connected to c_s holds a higher voltage $V(n_{c_s,high})$ than the other node $n_{c_s,low}$ connected to c_s. Starting at the node $n_{c_s,high}$ and $n_{c_s,low}$ the search algorithm analyses the network. In the following the search in the direction of $n_{c_s,low}$ is described. The analogous search in the direction of $n_{c_s,high}$ is indicated in brackets.

First all, cables $\mathcal{C}n_{c_s,low}$ ($\mathcal{C}n_{c_s,high}$) connected to $n_{c_s,low}$ ($n_{c_s,high}$) are checked for the strongest thermal overload and whether the voltage at the other end of the cable $V(n^*_{c_s,low})$ ($V(n^*_{c_s,high})$) is lower (higher) than the voltage at $n_{c_s,low}$ ($n_{c_s,high}$). The second condition ensures that the voltage along $\mathcal{S}_c$ monotonously decreases and guarantees a unidirectional current flow. The cable fulfilling both conditions is added to $\mathcal{S}_c$. Afterwards, the node $n^*_{c_s,low}$ ($n^*_{c_s,high}$) is set to the new $n_{c_s,low}$ ($n_{c_s,high}$). Starting at the new node, the process is repeated until no cable fulfilling the condition can be found 2 (3). This end point is either a local voltage minimum (maximum) and/or a transformer is connected there.

$$c_{s,low} = \underset{c_i \in \mathcal{C}n_{c_s,low}}{\arg\max} \left(\frac{I(c_i)}{I_{rated}(c_i)}\right) \wedge V(n^*_{c_s,low}) < V(n_{c_s,low}) \tag{2}$$

$$c_{s,high} = \underset{c_i \in \mathcal{C}n_{c_s,high}}{\arg\max} \left(\frac{I(c_i)}{I_{rated}(c_i)}\right) \wedge V(n^*_{c_s,high}) > V(n_{c_s,high}) \tag{3}$$

2.2.2. Find two nodes for a network extension cable Hereinafter $\mathcal{S}_c$ and F_{relief} are used to determine two nodes to which the new cable is added to the network. The search for these two nodes starts at V_{max} and V_{min} separately and analogously. Here we describe the search at V_{max}. The search beginning at V_{min} is indicated in brackets.

As indicated in equation 4 (5) out of all cables in $\mathcal{S}_c$ the cable $c_{s,V_{max}}$ ($c_{s,V_{min}}$) connected to the node with the highest (lowest) voltage is selected.

$$c_{s,V_{max}} = \underset{n_i \in \mathcal{S}_c}{\arg\max} \left(V(n_i)\right) \tag{4}$$

$$c_{s,V_{min}} = \underset{n_i \in \mathcal{S}_c}{\arg\min} \left(V(n_i)\right) \tag{5}$$

Finally, all cables which are an element of $\mathcal{S}_c$ are ordered by decreasing (increasing) nodal voltages. In a decreasing (increasing) order the cable c_s out of $\mathcal{S}_c$ is checked for condition 6 until the condition is fulfilled.

IOP Conf. Series: Journal of Physics: Conf. Series **977** (2018) 012006 doi:10.1088/1742-6596/977/1/012006

$$\left(\frac{I(c_s)}{I_{rated}(c_s)}\right) \geq \left(\frac{I(c_{I,max})}{I_{rated}(c_{I,max})}\right)(1 - F_{relief}) \tag{6}$$

If condition 6 is fulfilled, the connection point for a new cable is the node holding the higher (lower) voltage at c_s.

2.3. Network extension due to voltage range deviations

Here we describe how the algorithm reacts to a voltage range deviation. Based on the relief factor F_{relief} introduced in section 2 and the voltages at all nodes of the network $V(n_i)$, two nodes are found to which a new cable is added. The goal of the algorithm is to reduce voltage range deviations efficiently. First a sequence of nodes $\mathcal{S}_n$ with monotonously increasing voltages is found containing the node with the strongest voltage range violation. Afterwards the nodes are compared using F_{relief} to find two connection points for a new cable.

2.3.1. Find the node sequence with the highest voltage range deviation If a voltage range deviation occurs, the node $n_{\Delta V_{max}}$ in the network with the highest deviation from nominal voltage $|V(n_i) - V_{rated}|$ is found. Therefore, the node voltage $V(n_i)$ at all nodes in the network $\mathcal{N}$ is compared to the rated network voltage V_{rated}. Furthermore, the node with the highest voltage deviation $n_{\Delta V_{max}}$ is selected as n_s where the search algorithm starts (see equation 7).

$$n_s = n_{\Delta V_{max}} = \underset{n_i \in \mathcal{N}}{\arg\max}\,(|V(n_i) - V_{rated}|) \tag{7}$$

By definition $V(n_s)$ has to be the global maximum or minimum of the network voltage. Here we describe the case where n_s holds the minimum of the network voltage. The procedure in case that the voltage $V(n_s)$ is a global maximum is indicated in brackets. First the algorithm checks all cables $\mathcal{C}_n$ connected to n_s for the cable c_i where the voltage at the other end of $V(n_s^*)$ holds the highest (lowest) voltage. Additionally, the node n_s^* is selected as the new n_s and added to the sequence of nodes $\mathcal{S}_n$ (see equation 8 (9)).

$$n_s = \underset{c_i \in \mathcal{C}_n}{\arg\max}\,(V(n_s^*)) \wedge V(n_s^*) > V(n_s) \tag{8}$$

$$n_s = \underset{c_i \in \mathcal{C}_n}{\arg\min}\,(V(n_s^*)) \wedge V(n_s^*) < V(n_s) \tag{9}$$

Voltages at nodes are compared, selected and added to $\mathcal{S}_n$ until no cable is connected to n_s with a higher (lower) voltage at the other end n_s^* (see equation 10 (11) for the stopping criterion).

$$V(n_s) > \underset{c_i \in \mathcal{C}_n}{\arg\max}\,(V(n_s^*)) \tag{10}$$

$$V(n_s) < \underset{c_i \in \mathcal{C}_n}{\arg\min}\,(V(n_s^*)) \tag{11}$$

2.3.2. Find two nodes in the sequence of nodes to add a new cable between After $\mathcal{S}_n$ is complete, the algorithm searches for two nodes within $\mathcal{S}_n$ where a new cable is added to the network. First all nodes which are an element of $\mathcal{S}_n$ are compared by their voltage $V(n_i)$. The starting point for the search is that node n_s whose voltage is closed to the rated voltage V_{rated} of the network (Equation 12).

$$n_s = \underset{n_i \in \mathcal{S}_n}{\arg\min}\,(|V(n_i) - V_{rated}|) \tag{12}$$

Independently of whether a voltage range deviation is caused by a low or a high voltage, two searches are performed starting at n_s. One search is executed to nodes holding decreasing lower voltages than n_s, the other search leads towards nodes holding increasing voltages. Furthermore, all nodes have to be in $\mathcal{S}_n$ with the highest voltage. Here we describe the search towards the node with the highest voltage $n_{V_{max}}$. The analogous search in the direction of the node with the lowest voltage $n_{V_{min}}$ in $\mathcal{S}_n$ is provided in brackets.

The algorithm checks in an increasing (decreasing) order whether the local voltage $V(n_s)$ is higher (lower) than the nominal voltage plus (minus) an acceptable voltage deviation which depends on F_{relief} (see equation 13 (14)).

$$V(n_s) \geq V_{rated} + (F_{relief}\,(V(n_{V_{max}}) - V_{rated})) \tag{13}$$

$$V(n_s) \leq V_{rated} - (F_{relief}\,(V_{rated} - V(n_{V_{min}}))) \tag{14}$$

If equation 13 (14) is fulfilled, n_s is the node to which a new cable extends the network. If not, the node in $\mathcal{S}_n$ with the next higher (lower) voltage is selected. This process is repeated until a node has been found which fulfils equation 13 (14). After both nodes are found this way, the new cable connects these nodes along the shortest possible path.

3. Case study

Here we describe how the algorithm responds to voltage range deviations and thermal overloads on three networks using a given relief factor F_{relief}. We set up test cases for one radial, one loop, and one heavily meshed network (see section 3.1). Afterwards, the algorithm's response to these test cases is described (see section 3.2).

3.1. Test cases for a radial, a loop and a meshed network

The test case is defined for networks with three different topologies and a rated phase to phase voltage V_{rated} of 400 V. Each cable is of type NAYY-J 150 mm^2 with a maximal allowed one phase cable current of 275 A. The first topology is based on [18] and consists of two symmetric radial lines. Each line holds six nodes. The end of each line is connected to a slack node n_0 which represents a transformer (see Figure 1a). For the second topology we add an additional 60 m long cable between node n_6 and node n_{12} to the radial topology. This leads to the loop topology from Figure 1b. The third network is also based on the radial topology and is extended by five diagonal cables. These cables connect node n_1 with n_8, n_2 with n_9, n_3 with n_{10}, n_4 with n_{11} and n_5 with n_{12}. Each diagonal cable is 85 m long (Figure 1c).

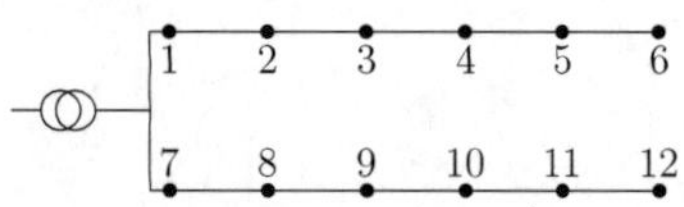

(a) Radial network

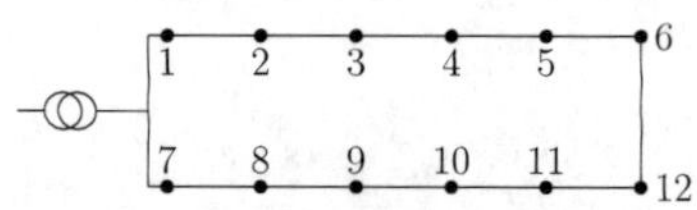

(b) Loop network

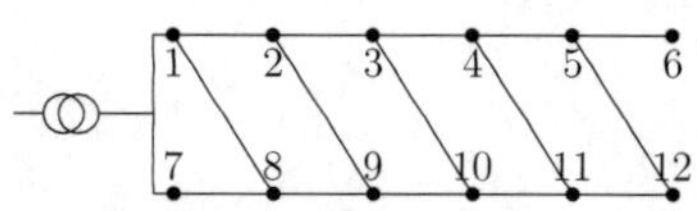

(c) Meshed network

Figure 1: Networks topologies

Here we allow a voltage range derivation of ±4 %. In order to test for the thermal response at each network every node n_1 to n_6 is connected to one 20 kW load. Each node from n_7 to n_{12} connects one 35 kW load. Afterwards, for the voltage range deviation response we reduce the load on node n_7 to n_{12} from 35 kW to 30 kW at each network. If an overload occurs, a new cable of type NAYY-J 150 mm^2 is added to the network. The release factor F_{relief} is set to 0.8 for the voltage range derivations and thermal overload response. The network parameters are summarized in table 1.

Table 1: Parameters for the radial, the loop and the meshed network topology

Transformer (slack node)	n_0
No. of nodes	$12 + n_0$
Cable type	NAYY-J 150 mm^2
Cable length between nodes:	
Radial grid	60 m
Closing calbe for loop grid	60 m
Diagonal cables at meshed grid	85 m
Load at n_1 to n_6	20 kW ($\cos(\phi) = 1$)
Thermal response:	
Load at n_7 to n_{12}	35 kW ($\cos(\phi) = 1$)
Voltage response:	
Load at n_7 to n_{12}	30 kW ($\cos(\phi) = 1$)

3.2. Results

3.2.1. Thermal overload response In the case that 35 kW loads are connected at n_1 to n_7 the thermal overload response triggers at all three networks. For the radial network the cable with the strongest thermal overload $c_{I,max}$ is the cable between node n_0 and n_7. The current on that cable $|I(c_{I,max})|$ is 318 A and therefore higher than the rated current of 275 A of that cable. The extension algorithm responds to the overload by placing a new cable between n_0 and n_9. This reduces the maximal current on that cable by 43 %. At the loop network the strongest overloaded cable is also the cable between node n_0 and n_7. Here the current is 278 A before and 185 A after the extension algorithm responses. In this case a new cable is added between n_0 and n_9. The cable with the strongest thermal overload at the meshed grid is the cable between n_0 and n_1. The current on that cable is 286 A before the algorithm for thermal overloads responds. The algorithm builds a new cable between n_0 and n_2. This reduces the current at the cable between n_0 and n_1 by 35 % (see table 2).

Table 2: Electric currents of cables with the strongest thermal overload $|I(c_{I,max})|$ and cables by which the networks are extended.

	$c_{I,max}$		$\|I(c_{I,max})\|$		Extended cable	
Network	n_{from}	n_{to}	Before ext.	After ext.	n_{from}	n_{to}
Radial	n_0	n_7	318 A	182 A	n_0	n_9
Loop	n_0	n_7	278 A	185 A	n_0	n_9
Meshed	n_0	n_1	286 A	187 A	n_0	n_2

3.2.2. Voltage range deviation response If the load at all nodes from n_1 to n_7 is reduced from 35 kW to 30 kW, the voltage range deviation response triggers at all networks. The highest voltage drop at the radial network occurs at node $n_{12} = n_{\Delta V_{max}}$. At that node the voltage drops by 5 % to 380 V. At the radial network the algorithm reacts by placing a new cable between node n_0 and n_8. After the extension the maximal voltage drops to 385 V occurs at node n_{12}. For the loop network the node with the highest voltage deviation $n_{\Delta V_{max}}$ is also node n_{12}. The voltage drops at this node to 383 V. Therefore, the algorithm extends the network by a new cable between node n_0 and n_8. This leads to a maximal voltage deviation at node n_{12} with a voltage drop of 3.5 %. At the meshed network node n_6 is the node with the highest voltage deviation $n_{\Delta V_{max}}$. At this network the highest voltage drop occurs at node n_6 by 4.3 % before the network extension. The algorithm responds by adding a new cable between node n_0 and n_2. After this extension the highest voltage deviation occurs at node n_6 with 387 V (see table 3).

Table 3: Voltages at the nodes with the highest voltage deviation $|V(n_{\Delta V_{max}})|$ before and after the voltage extension response for the radial, the loop and the meshed network.

	Before ext.		After ext.		Extended cable	
Network	$n_{\Delta V_{max}}$	$\|V(n_{\Delta V_{max}})\|$	$n_{\Delta V_{max}}$	$\|V(n_{\Delta V_{max}})\|$	n_{from}	n_{to}
Radial	n_{12}	380 V	n_{12}	385 V	n_0	n_8
Loop	n_{12}	383 V	n_{12}	386 V	n_0	n_8
Meshed	n_6	383 V	n_6	387 V	n_0	n_2

4. Conclusions

This paper introduces an extension algorithm which reacts to critical network states by placing additional cables to an existing network. The algorithm requires the network topology, the current and maximal current of every cable as well as the voltages at all nodes of the network. Furthermore, a maximal voltage deviation range and a release factor must be set. Here we test the algorithm in a case study with a radial, a loop, and a meshed electrical network.[2] The algorithm distinguishes between voltage range deviations and thermal overloads. We found that the algorithm functions for each of these networks adequately and reliably. For all examined networks the algorithm is capable of eliminating thermal overloads and voltage drops by adding one cable to the given networks. Based on these findings it can be concluded that the suggested algorithm offers an easy-to-use approach to estimate investments for electric low voltage networks. Especially for generic low voltage networks where just the network topology is available, the algorithm is capable of estimating network extension needs. Compared to methods for optimal network planning, the algorithm does not need information about routes for additional cables or positions for additional substations. Due to the algorithm design, adding additional cables is the only response for network extensions. Additional extension options are not taken into consideration. Furthermore, no forecast for additional loads and their placement are taken into consideration when placing additional cables. Therefore, the main value of the work is that the algorithm is a flexible and an easy-to-use approach for calculating grid extension costs for large scale network costs assessments. The algorithm works sufficiently to estimate investment needs especially on generic networks of which no detailed information is available but the algorithm is not a tool for detailed network planning.

Acknowledgments
This work has been financed by the Federal Ministry for Economic Affairs and Energy within the project "Flexible Nachfrage als wichtiger Beitrag zur Energiewende und Baustein in der Energiesystemanalyse EnSYS-FlexA". We thank the members for their support. We also thank Barbara Sinnemann and the reviewers for critically reading the manuscript.

[2] The algorithm works well for more complex networks. Networks of any topology can be set up and examined as JUnit tests. These tests are available at the project homepage.[1]

IOP Conf. Series: Journal of Physics: Conf. Series **977** (2018) 012006 doi:10.1088/1742-6596/977/1/012006

References

[1] R. N. Adams and M. A. Laughton. Optimal planning of power networks using mixed-integer programming. Part 1: Static and time-phased network synthesis. *Proceedings of the Institution of Electrical Engineers*, 121(2):139, 1974.

[2] Annegret-Cl. Agricola, Bernd Höflich, Phillipp Richard, Jakob Völker, Christian Rehtanz, Marco Greve, Björn Gwisdorf, Jan Kays, Noll Theresa, Johannes Schwippe, André Seack, Jan Teuwsen, Gert Prof. Dr. Brunekreeft, Roland Dr. Meyer, and Vanessa Dr. Liebert. Ausbau- und Innovationsbedarf in den Stromverteilnetzen in Deutschland bis 2030: dena-Verteilnetzstudie, 2012.

[3] BAFA. Bundesamt für Wirtschaft und Ausfuhrkontrolle: Elektromobilität (Umweltbonus), 2016.

[4] BMWi. Stromnetze der Zukunft, 2016.

[5] BNetzA. Bundesnetzagentur - Übersicht Strom- und Gasnetzbetreiber, 2016.

[6] Jens Büchner, Jörg Katzfey, Ole Flörcken, Albert Moser, Henning Schuster, Sebastian Dierkes, Tobias van Leeuwen, Lukas Verheggen, Mathias Uslar, and Marie van Amelsvoort. Moderne Verteilernetze für Deutschland (Verteilernetzstudie): Abschlussbericht, 2014.

[7] Bundestag. Gesetz zur Einführung von Ausschreibungen für Strom aus erneuerbaren Energien und zu weiteren Änderungen des Rechts der erneuerbaren Energien: Erneuerbare-Energien-Gesetz – EEG 2016, 08.07.16.

[8] Till Gnann, Patrick Plötz, Simon Funke, and Martin Wietschel. What is the market potential of plug-in electric vehicles as commercial passenger cars? A case study from Germany. *Transportation Research Part D: Transport and Environment*, 37:171–187, 2015.

[9] J.F Gomez, H.M Khodr, P.M de Oliveira, L. Ocque, J. M. Yusta, R. Villasana, and A.J Urdaneta. Ant colony system algorithm for the planning of primary distribution circuits. *Power Systems, IEEE Transactions on*, 19(2):996–1004, 2004.

[10] T. Gönen and B. L. Foote. Distribution-system planning using mixed-integer programming. *IEE Proceedings C Generation, Transmission and Distribution*, 128(2):70, 1981.

[11] T. Gönen and I. J. Ramirez-Rosado. Review of distribution system planning models: a model for optimal multistage planning. *IEE Proceedings C - Generation, Transmission and Distribution*, 133(7):397–408, 1986.

[12] M.O.W. Grond, J. Morren, and H. Slootweg. Requirements for advanced decision support tools in future distribution network planning. In *CIRED*, page 1046, [London] and [Stevenage], 2013. [Institution of Engineering and Technology] and [CIRED].

[13] Gunther Gust, Wolfgang Biener, Tobias Brandt, Kilian Dallmer-Zerbe, Dirk Neumann, and Bernhard Wille-Haussmann. Decision support for distribution grid planning. In *2016 IEEE International Energy Conference (ENERGYCON)*, pages 1–7. IEEE, 2016.

[14] S. K. Khator and L. C. Leung. Power distribution planning: A review of models and issues. *IEEE Transactions on Power Systems*, 12(3):1151–1159, 1997.

[15] Simon Marwitz, Marian Klobasa, and David Dallinger. Comparison of control strategies for electric vehicles on a low voltage level electrical distribution grid. In *International Symposium on Energy System Optimization*. Heidelberg, 2015.

[16] P. C. Paiva, H. M. Khodr, J. A. Dominguez-Navarro, J. M. Yusta, and A. J. Urdaneta. Integral Planning of Primary–Secondary Distribution Systems Using Mixed Integer Linear Programming. *IEEE Transactions on Power Systems*, 20(2):1134–1143, 2005.

[17] Luis Pieltain Fernandez, Gomez San Roman, Tomás, Rafael Cossent, Carlos Mateo Domingo, and Pablo Frias. Assessment of the Impact of Plug-in Electric Vehicles on Distribution Networks. *IEEE Transactions on Power Systems*, 26(1):206–213, 2011.

[18] Yves-Marie Saint-Drenan, Amany von Oehsen, Thomas Stetz, and Martin Braun. Vorstudie zur Integration großer Anteile Photovoltaik in die elektrische Energieversorgung: Studie im Auftrag des BSW - Bundesverband Solarwirtschaft e.V. Ergänzte Fassung vom 29.05.2012, 2012.

[19] Remco A. Verzijlbergh, Marinus O. W. Grond, Zofia Lukszo, Johannes G. Slootweg, and Marija D. Ilic. Network Impacts and Cost Savings of Controlled EV Charging. *IEEE Transactions on Smart Grid*, 3(3):1203–1212, 2012.

[20] J. von Appen, M. Braun, T. Stetz, K. Diwold, and D. Geibel. Time in the Sun: The Challenge of High PV Penetration in the German Electric Grid. *Power and Energy Magazine, IEEE*, 11(2):55–64, 2013.

[21] H. Willis and J. D. Northcote-green. Comparison of Several Computerized Distribution Planning Methods. *IEEE Transactions on Power Apparatus and Systems*, PAS-104(1):233–240, 1985.

[22] Jenny Winker, Philipp Oehler, Anna-Lena Klingler, Frank Sensfuß, Holger Höfling, Mario Ragwitz, Bernd Tersteegen, Maurer Christoph, Christian Linke, Nick Seeger, Alexander Ladermann, Dominik Greinacher, Günther Reinald, Henning Jachmann, and Frank Musiol. Zukunftswerkstatt Erneuerbare Energien: Eigenversorgung mit Strom Analyse der Wirkungen und Szenarien für die zukünftige Entwicklung

(Endbericht): Studie im Auftrag des Bundesministeriums für, 2016.
[23] Rana.H.A. Zubo, Geev Mokryani, Haile-Selassie Rajamani, Jamshid Aghaei, Taher Niknam, and Prashant Pillai. Operation and planning of distribution networks with integration of renewable distributed generators considering uncertainties: A review. *Renewable and Sustainable Energy Reviews*, 2016.

IOP Conf. Series: Journal of Physics: Conf. Series **977** (2018) 012007 doi:10.1088/1742-6596/977/1/012007

The eGo grid model: An open-source and open-data based synthetic medium-voltage grid model for distribution power supply systems

J Amme[1], G Pleßmann[1], J Bühler[1], L Hülk[1], E Kötter[1] and P Schwaegerl[2]

[1] Reiner Lemoine Institut gGmbH, Rudower Chaussee 12, 12489 Berlin, Germany
[2] LEW Verteilnetz GmbH, Schaezlerstraße 3, 86150 Augsburg, Germany

E-mail: `jonathan.amme@rl-institut.de`

Abstract. The increasing integration of renewable energy into the electricity supply system creates new challenges for distribution grids. The planning and operation of distribution systems requires appropriate grid models that consider the heterogeneity of existing grids.

In this paper, we describe a novel method to generate synthetic medium-voltage (MV) grids, which we applied in our DIstribution Network GeneratOr (DINGO). DINGO is open-source software and uses freely available data.

Medium-voltage grid topologies are synthesized based on location and electricity demand in defined demand areas. For this purpose, we use GIS data containing demand areas with high-resolution spatial data on physical properties, land use, energy, and demography. The grid topology is treated as a capacitated vehicle routing problem (CVRP) combined with a *local search* metaheuristics. We also consider the current planning principles for MV distribution networks, paying special attention to line congestion and voltage limit violations. In the modelling process, we included power flow calculations for validation.

The resulting grid model datasets contain 3608 synthetic MV grids in high resolution, covering all of Germany and taking local characteristics into account. We compared the modelled networks with real network data. In terms of number of transformers and total cable length, we conclude that the method presented in this paper generates realistic grids that could be used to implement a cost-optimised electrical energy system.

1. Introduction

The transition of the energy system in Germany from fossil fuels to renewable energy sources (RES) (the *Energiewende*) brings new challenges to grid operation and extension planning. These include new demands on grid topology, vast grid expansion needs, and reverse power flows in distribution grids (DGs) due to distributed generation in low and medium-voltage grids. The first point is addressed by the *Netzentwicklungsplan* (NEP), a public participatory process to identify the expansion needs of the transmission grid [1]. However, the NEP has shortcomings. Specifically these are considering the DG extension planning while determining transmission grid extension needs and a severe incorporation of batteries as one flexibility option. Separated grid extension planning of the transmission and distribution grid level [2, 3] may overestimate overall grid extension needs in terms of capacity and costs. The research project *open_eGo* [4] tackles this problem by interlinking grid extension planning and operation of transmission and distribution systems. Our objective is to unveil untapped cost reduction potential of combined planning across the transmission and distribution grid levels. To accomplish this, our method considers the requirements of the transmission system in planning distribution grids. The work described in this paper is complemented by other work of Müller et al. on the eGo transmission grid model [5].

Our research requires access to grid topology data and assets such as transformers. Whereas grid data at the transmission level are available through *Open Street Map* (OSM) [6], publicly accessible data of DGs are missing. Generally, distribution system operators (DSOs) do not publish infrastructure data. This hinders research on grid integration of RES. How might this problem be addressed? The *Conseil International des Grands Réseaux Électriques* (CIGRE) provides sample medium-voltage (MV) grid data having typical characteristics of MV grids in Europe [7]. Operators do publish some structural characteristics of grids in order to comply with German legislation [8, 9]. One study analyzing DGs used real grid data provided by operators to partners in a specific project [2]. To obtain a variety of distinct MV and low-voltage (LV) grid data, Monte-Carlo variations of structural characteristics have been used [3]. Rui et al. created synthetic MV test grids on the basis of two-dimensional combinations of load density [10]. Some define reference grids which approximate a specific set of real-world grids [11].

So far, however, there are no freely available tools or methods for modelling the full variety of DGs found in Germany while taking into account local characteristics such as loads and generators. We wanted to develop a set of methods that generate a representative dataset of DGs in Germany using open data. The methods we developed produce synthetic MV grids for Germany. We use high-resolution spatial data of load and generation in order to approximate the grid topology. The approximation of MV grid topology is achieved using a routing technique [12, 13]. The methods, model parameters, and assumptions used in this work were developed in close collaboration with grid planning experts from the German DSO LVN. Our paper focuses on modelling MV grid topology; the construction of LV grids will be treated in a future publication. Low-voltage grids are represented by aggregated load and generation at MV-LV substations. The implementation uses open data and is released under an open-source licence.

2. Fundamental data basis

Our objective is to synthesize status quo MV grid topologies. For this reason, we use an approach to reflect their historic development. Historically, DGs were created with the goal to distribute power from large central power plants to customers. Thus, the peak load was the crucial design parameter [14]. However, as capacity of RES increases, the focus on peak demand for DG planning will become less important [15]. In order to reflect grid topology that was originally built decades ago, we focus on peak demand as the main parameter. The impact of RES generation units on grids will be subsequently considered.

To synthesize a MV grid topology based on the peak load, a high-resolution spatial representation of the electricity demand is required. This is provided by Hülk et al. [16] and is available on the *OpenEnergy Platform* (OEP) [17], which was developed within the *open_eGo* research project. The data model of Hülk et al. (see Fig. 1) includes, among other data, the so-called medium-voltage grid districts (MVGDs) which are supply areas of MV grids named MVGDs. These typically contain multiple load areas (LAs), which are defined as geographic clusters where electricity is consumed. Sector-specific electricity demand are associated to these LAs. The LAs are identified by a GIS analysis of high-resolution spatial data which contain information about physical properties, land use, energy, and demography.

Within each LA, we define locations of MV-LV substations using an equidistant grid of points with an interval of 360 m. Each of these substations has a supply area called low-voltage grid district (LVGD). The LVGDs are defined by a Voronoi partition of a LA using the locations of the MV-LV substations. The result is a hierarchical representation of real supply structures.

IOP Conf. Series: Journal of Physics: Conf. Series **977** (2018) 012007 doi:10.1088/1742-6596/977/1/012007

Figure 2 perfectly visualizes this structure of a MVGD with its areas and substations.

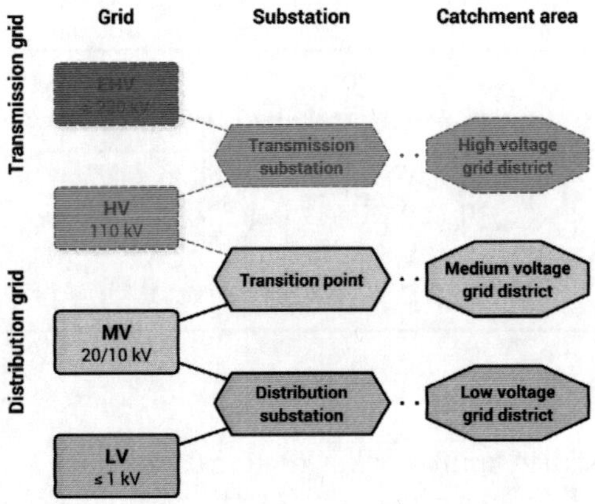

Figure 1: Simplified scheme of voltage levels, substations, and areas. Figure adapted from [16]).

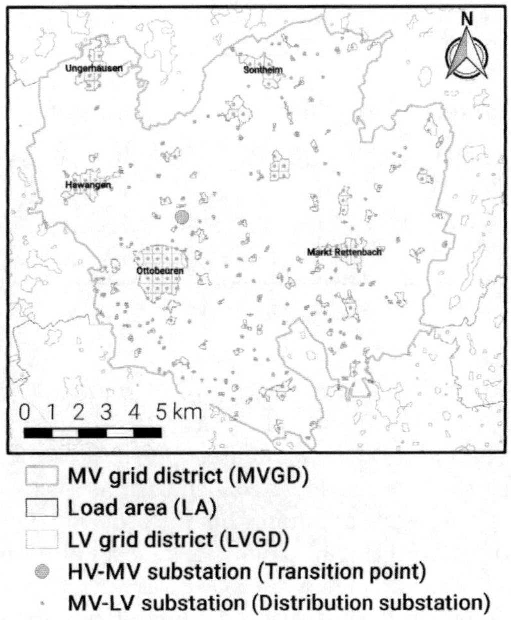

Figure 2: Example medium-voltage grid district (MVGD) illustrating the spatial data model.

Annual electricity consumption is assigned to LAs for the following sectors: residential, retail, industrial, and agricultural [5]. Using standard load profiles for Germany [18], we obtain the peak load for each of these areas. The standard load profiles are supplemented by a profile describing industrial electricity demand assuming discrete load levels (weekday 6am-10pm: 0.8, other 0.6). It is scaled by actual annual consumption.

Data on conventional and renewable generators are publicly available [19, 20]. We use preprocessed versions of these datasets from [21, 22]. Some of them contain implausible information on grid level, therefore generators are allocated to grid levels according to Table 1 [16]. A realistic DG model requires power plants with precise information about their voltage levels and geographic locations. Conventional power plants are provided with exact locations, whereas RES data have a low spatial accuracy. Due to privacy concerns, its spatial resolution is reduced by the original data provider. This leads to errors of up to several kilometres. In order to obtain realistic locations, we allocate RES capacity to certain areas. The targeted allocation areas are defined for each type and subtype of each technology as presented in Table 2. This is achieved by using different technology-specific distribution methods as described below.

Biomass and gas power plants are relocated to agricultural areas of MVGDs presuming a spatial proximity to the origin of their fuel. The same areas are used for roof-mounted solar plants of voltage levels 4 and 5. We assume these are usually located on farm buildings that typically provide suitable sites for medium-size solar plants, whereas ground-mounted solar generators are distributed in the MVGDs to random locations. The data contain wind energy converters (WECs) of voltage levels 4 and 5 [20]. Those of voltage level 4 are expected to be wind farms or clusters of single WECs. They are relocated to wind potential areas (WPAs) provided by [23]. WPAs are defined as sites that are not classified as unsuitable for use of WECs (i.e. settlement areas, national parks, and infrastructure) and that are sufficiently distant

Table 1: Allocation of generators to voltage level according to their nominal capacity, sum of allocated capacity for Germany (adapted from [16]).

Voltage level	**Nominal capacity**	**Allocation target**	**Cumulative cap. (renewable)**	**Cumulative cap. (conv.)**
4 (HV-MV)	4.5 - 17.5 MW	Transition point	9.73 GW	2.48 GW
5 (MV)	0.3 - 4.5 MW	MV grid	34.89 GW	0.05 GW
6 (MV-LV)	0.1 - 0.3 MW	Distribution subst.	5.18 GW	0 GW
7 (LV)	$\leq$ 0.1 MW	LV grid	21.33 GW	0 GW

from settlements in order to be compliant with distance limits. The reallocation takes place successively in descending order in terms of nominal capacity and size of the WPA. Voltage levels 5 and 6 are dominated by single scattered WECs which are distributed throughout the WPAs. To avoid local concentrations, WPAs are represented by a $500 \times 500\,\mathrm{m}^2$ grid of points that ensures a minimum distance of 500 m between two WECs, which is a common value for current WECs [24]. Hydroelectric and geothermal power plants retain their original location. Finally, solar and wind energy plants of voltage level 7 are assigned to LVGDs. These are assumed to be installed on roofs of buildings (solar plants) or nearby buildings (WEC). The distance between buildings varies considerably, several analyses show a typical range between 9 m and 102 m [25, 26, 27]. We assume an average distance of 50 m. Therefore, we derive potential locations by applying a 50 x 50 m grid of points on each LVGD.

The processed data form a high-resolution spatial model of load and generation in Germany which is used to determine distribution grids as described below.

3. Synthetic medium-voltage grids

The structure of synthetic MV grids resulting from the application of our method depends on the choice of input data and the assumptions, which we will review below.

3.1. Types of medium-voltage grid districts and load areas

For historical reasons and due to local conditions, MV grids have certain topologies. The load and the size of a supply area were crucial parameters of grid design [28]. Meshed grids are mainly used in urban areas with high load density, whereas radial and ring structures are typically used in regions of low load density. Since 84.3 % of German MV grids have a ring topology [3], we assume this type of topology for all MV grids analysed in our paper. This topology consists of a main route and branching stubs and provides a reasonable trade-off between costs and reliability of supply [12].

First, we categorise MVGDs as urban or rural supply areas. Urban areas are usually equipped with underground cables at a voltage of 10 kV, whereas grids in rural areas mostly constitute of overhead lines at a rated voltage of 20 kV [28, 12]. We use the maximum distance from the HV-MV substation to the LAs' centres and a value of 1 kV/km line length [28]. To distinguish between these two area types, we define a threshold of 15 km. Consequently, MVGDs with a maximum distance of $\geq$ 15 km are treated as rural (20 kV), whereas MVGDs with $<$ 15 km are treated as urban areas (10 kV).

In accordance with these prerequisites, the LAs (introduced in Section 2) are categorised according to their peak load. We define three categories of LAs (see Figure 3):

(1) *Aggregated* LAs represent regions with high cumulative peak load. Future grid problems

IOP Conf. Series: Journal of Physics: Conf. Series **977** (2018) 012007 doi:10.1088/1742-6596/977/1/012007

Table 2: Spatial distribution of generators according to type and sub-type [20] throughout distribution supply area.

Voltage level	Generation type	Generation subtype	Allocation area	Cumulative cap. [MW]
4 (HV-MV)	biomass, gas	all	agricultural area	943
	geothermal	all	-	27
	hydro	all	-	235
	solar	roof-mounted	agricultural area	156
	solar	ground-mounted, none	randomly in MVGD	3923
	wind	all	WPA (wind farms)	4451
5 (MV)	biomass, gas	all	agricultural area	4835
	geothermal	all	-	12
	hydro	all	-	957
	solar	roof-mounted	agricultural area	2785
	solar	ground-mounted, none	randomly in MVGD	5893
	wind	all	WPA	20408
6 (MV-LV)	biomass, gas	all	agricultural area	1029
	hydro	all	-	162
	solar	all	randomly in LVGDs	3942
	wind	all	WPA	42
7 (LV)	biomass, gas	all	agricultural area	128
	hydro	all	-	164
	solar	all	randomly in LVGDs	21011
	wind	all	randomly in LVGDs	24

involve line congestion and violation of the maximum permissible voltage band produced by generators. They will rarely occur in urban areas because these areas have fewer available RES sites and comparatively strong electrical grids to integrate them. Therefore, regions of high peak load are excluded from grid reinforcement measures and modelled using aggregated load and generation capacities. The threshold is defined by the current carrying capacity of the line used to connect *Regular* LAs. *Aggregated* LAs are directly connected to the MV bus bar of the HV-MV substation.

(2) *Regular* LAs are connected to MV rings and thus determine the rings' topology. As the literature does not distinguish between LAs that are part of the main ring and those connected via stubs, we assume a minimum peak load of 100 kW for LAs being considered in the main route of a ring (this assumption is made based on personal communication with the grid planning department of a DSO [29]).

(3) *Satellite* LAs incorporate regions with low electricity peak demand ($\leq$ 100 kVA) and lower requirements in terms of reliability of supply. These are connected to rings via branch lines. LAs with a peak load of less than 1 kVA are assumed to be the result of errors in data processing and therefore omitted.

3.2. Technical constraints and scenarios

When planning and operating DGs, DSOs have to comply with the legal framework; The system must meet minimum technical requirements. In Germany, MV grids must be able to withstand

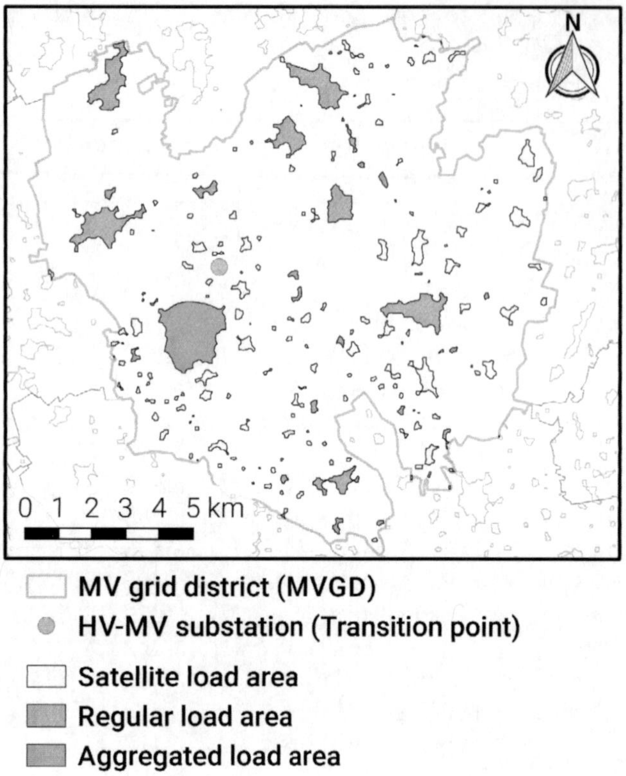

Figure 3: Types of load areas.

the outage of any single component e.g. a cable or transformer ((n-1)-criterion). Redundancy is realised using switch disconnectors in MV-LV substations to isolate parts of the grid in case of a failure. HV-MV substations meet this requirement by using at least two parallel transformers. Common practice of DSOs is to evaluate two worst-case scenarios in order to test the stability of grids [30]: heavy load flow (annual peak load, min. generation) and reverse power flow (max. generation, min. load). The most important aspects of grid stability in this context are thermal ampacity of lines and transformers, and the tolerable voltage range. The (n-1)-criterion applies to loads only. Generators connected to the grid are not (n-1)-secure. To ensure (n-1)-secure supply, transformers and lines need to have reserve capacity. Therefore, different load factors for equipment are used in power flow analysis (see Table 3 [2]). The (n-1)-criterion does not apply to LV networks.

Table 3: Load factors of equipment (relative to rated apparent power) [2]

Equipment	**Load factor (LF) heavy load flow**	**Load factor reverse power flow**
HV-MV transformer	max. 60 %	max. 100 %
MV cable	max. 60 %	max. 100 %
MV-LV transformer	max. 100 %	max. 100 %
LV cable/overhead line	max. 100 %	max. 100 %

To account for temporal variability of load and generation, typically the coincidence factor is used. It reflects the coincidence of a group of loads and the availability of generation of a group of generators [2]. As our method assumes the minimum generation/load with zero and the maximum generation/load with 1, these factors have no effect and are neglected.

Furthermore, constraints on voltage stability apply. The EN 50160 standard defines the acceptable voltage range that must be guaranteed for each customer connected to the grid [31]. Grid codes define rules for connection of RES in MV grids [18] and LV grids [32]. To account for reactive power, we assume a power factor $\cos\varphi$ of 0.9 [12] for all loads and 1.0 for generators. Our grids are equipped with standard cable types, overhead lines, and transformers according to [33, 28, 12] (see Table A1 in the Appendix). MV-LV substations are equipped with ideal transformers neglecting losses.

3.3. Grid topology

As mentioned above, an open ring structure of the MV grids was chosen as the base topology. The LAs from Section 3.1 are supplied by an indefinite number of rings, which are fed by a single HV-MV substation. The identification of MV rings is based on a method from operations research – the capacitated vehicle routing problem (CVRP) [34]. In the CVRP, customers are supplied by vehicles with limited capacity that start their route from a central depot. In an electrical grid, the HV-MV substation serves as depot while customers are represented by the centers of load areas. The center of an LA is calculated as the centroid or as a point on the surface if the centroid is located outside of the area. The objective is to minimise the cumulative route length. In our case, material and installation of lines are the main cost factors [35]. This is based on fundamental work by Tao [12] who applied CVRP to grid planning. There is a variety of approaches to solve CVRPs including exact, heuristic, and metaheuristic techniques [36]. We employ a two-stage metaheuristics. The initial route construction is done using the parallel *savings* heuristic of Clarke and Wright [37]. It is subsequently improved by a *local search* using *graph operators* as described in Section 3.3.2. Figure 4 presents steps of MV grid generation at a glance. The improved topology of the MV grid's main route is followed by extending the topology with additional dispersed LAs and generators. In the last two steps the resulting grid model is tested for grid stability and reinforced if required.

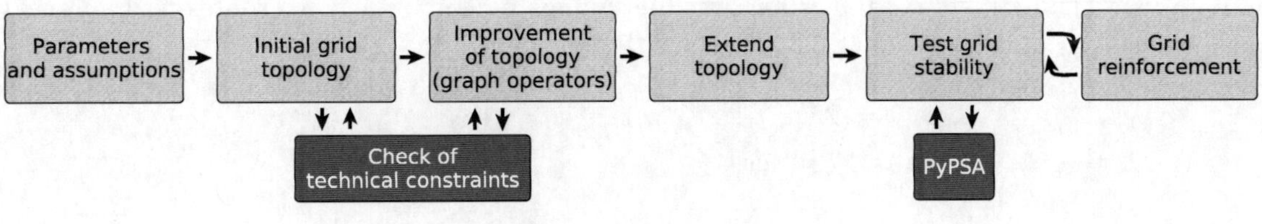

Figure 4: Steps of MV grid generation.

In reality, routes of underground cables and overhead lines follow geographical characteristics such as transport infrastructure [38]. Since the approach does not involve these constraints, the estimated route lengths tend to be underestimated. To account for differences in route length, a detour factor is used. Typically this value ranges from 1.1 to 1.5 [12]. We assume a detour factor of 1.3.

3.3.1. Identification of base topology

The main route of a MV grid is identified by formulating a CVRP while considering technical constraints (cf. Section 3.2). In order to find the initial solution of the CVRP the maximum rated apparent power $S_{line,max,th}$ of lines and cables in each MVGD must be defined. This depends primarily on the cumulative peak load $S_{MVGD,max}$ of all *regular* LAs (cf. Section 3.1).

Therefore, we assume the typical number of outgoing lines/half-rings $n_{lines} = 8$ for each MVGD [3]. The line type is selected such that

$$S_{line,max,th} \geq \frac{S_{MVGD,max} \cdot LF}{n_{lines}} \tag{1}$$

with LF referring to the load factor in the heavy load flow case. In doing so, we suppose an equally distributed load for each of the half-rings as reflected by Equation (1). As line and cable equipment has discrete rated power $S_{line,max,th}$, we always choose the type of the next larger value (see Table A1 in the Appendix). This identified line type is used as a default for all lines in the particular MVGD. The HV-MV substation is equipped with two transformers each having a nominal apparent power greater than $S_{MVGD,max} \cdot LF$ (as small as possible). Their operating voltage at the MV bus bar is set to nominal voltage.

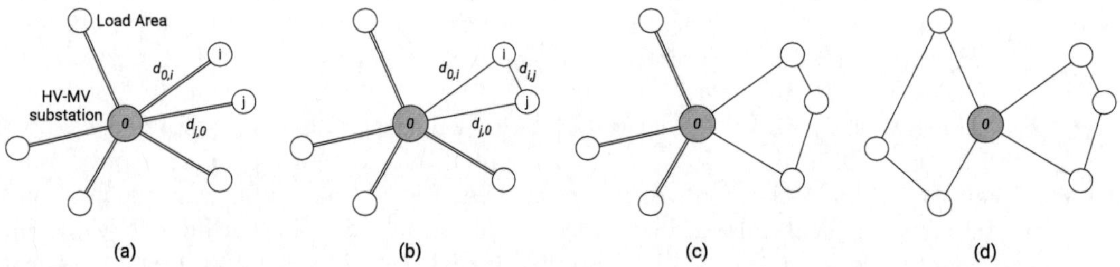

Figure 5: Example for *savings* heuristic: (a) initial solution, (b,c) iterations, (d) savings solution. Figure adapted from [12].

Using these technical guidelines, the *savings* heuristic can be applied to CVRP. The grid structure obtained employing this method is described by a graph structure. Lines are represented by edges, while loads and generators are nodes in the graph. In the initial solution, each load area is located on a separate route which consists of two edges, see the example in Figure 5 (a). The next step searches for possible savings by combining two routes with one-way distances i.e. $d_{0,i}$ and $d_{j,0}$. Equation (2) shows the saving $s_{i,j}$ for one step ((b) of Figure 5).

$$\begin{aligned} s_{i,j} &= (2 \cdot d_{0,i} + 2 \cdot d_{j,0}) - (d_{0,i} + d_{i,j} + d_{j,0}) \\ &= d_{0,i} + d_{0,j} - d_{i,j} \end{aligned} \tag{2}$$

The savings are calculated for all combinations of two routes and subsequently sorted in descending order by the amount of savings. These savings are successively applied (b,c) until there are no further improvements left (d).

The feasibility of a move is determined by restrictions on current carrying capacity and voltage stability. They are continuously verified in each iteration (cf. Section 3.2). Therefore, we use two modes of operation (normal and faulty) and adjust thresholds of tolerable current and voltage:

(1) Under normal operating conditions, MV rings are operated as isolated half-rings [28]. These are usually split by a switch disconnector at the location creating minimal power flows in closed conditions [38]. Subsequently, current carrying capacity and voltage stability are checked for each of the two half-rings as depicted in Figure 6 (a). A maximum line loading of 60 % [2] and a maximum voltage drop of 5 % [15] from the secondary side of the HV-MV substation to the primary side of the MV-LV substations are accepted.

(2) In case of a failure, the faulty segment is isolated by switch disconnectors, while the remaining part of the ring is fed by one side (b). In the worst case, the faulty segment is the closest to the substation. For this purpose, we allow a maximum line loading of 100 % of its current rating [2] and a tolerable voltage drop of 10 % [39].

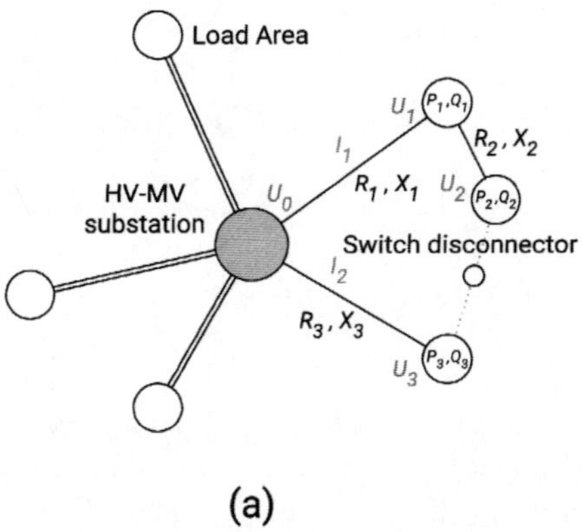

Figure 6: Modes of operation: (a) normal, (b) failure in one half-ring. (For visualisation purposes, the switch disconnector is located on line).

Due to the high number of route alterations and computational cost of power flow calculations, simplified methods for detecting line congestion and voltage violations are required. Current rating violations can be detected with small computational effort using the cumulative peak load of a ring. The total voltage drop ΔU, however, needs to be calculated for every node considering active and reactive power. In terms of the longitudinal voltage, this can be done by using Equation (3) assuming the transverse voltage drop to be negligible [18, 40].

$$\Delta U = \sum_{i=1}^{n} \Delta U_i = \sum_{i=1}^{n} \frac{(R_k \cdot P_i + X_k \cdot Q_i)}{U_N}, \tag{3}$$

R_k being the total resistance and X_k being the total reactance of lines from node $i = \{1, ..., n\}$ to the MV-LV substation. U_N equals the nominal voltage of the grid.

Moreover, we use additional criteria to decide on the feasibility of a move:

- The maximum number of substations per ring is 20 [28].
- The maximum length of a full ring is 60 km [38].

Following this method, computable and technically stable MV rings are generated. However, the *savings* heuristic produces solutions which are not necessarily optimal [34].

3.3.2. Improvement of initial topology

The identified grid topology is iteratively improved by exploring neighbouring solutions using a *local search* heuristic. We apply so-called *graph operators* to reduce the cumulative length of all rings in a MVGD. These include intra-route and inter-route techniques [41]. Intra-route operators apply modifications on a single route whereas inter-route operators involve segments of two routes.

The Or-Opt operator (intra-route) relocates all possible chains of three to one consecutive nodes within a route as illustrated in Figure 7 (a). It starts with three nodes followed by chains

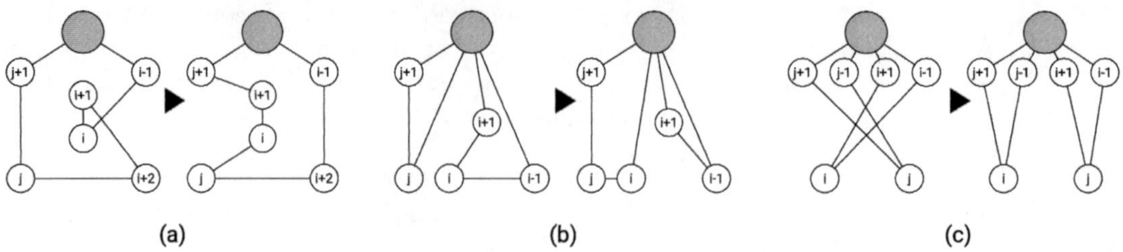

Figure 7: Graph operators: (a) Or-Opt (intra-route), (b) Relocate (inter-route), (c) Exchange (inter-route).

with two nodes and finally relocates single nodes. The relocate operator (inter-route) moves a single node from one route to another (b). The exchange heuristic swaps two nodes of different routes (c). The three methods are applied consecutively in the order as mentioned. During execution of every method, the best of all possible moves is performed starting over again until there are no feasible improvements. The acceptance criterion given by the technical constraints (cf. Section 3.2) is tested for every operation on the graph.

The modifications can reduce the cumulative ring length in a MVGD. The potential for reduction by *graph operators* depends heavily on the quality of the initial *savings* solution [36]. It may happen that initial solutions are very close to a (local) optimum which cannot be improved by *graph operators*.

3.3.3. Extending the base topology

The initial MV grid topology omits connection of *satellite* LAs, generation units, and MV-LV substations in the LAs. These are connected to the ring after identification of the MV grid main route. Therefore, different grid connection principles are applied in a particular order:

(1) Nodes close to an existing route of the grid ($\leq$ 100 m) are included by modifying the route's course slightly, making the node part of the route [29] (Figure 8 (1)). This applies not only to ring routes' lines but also to branch lines.

(2) Remaining nodes are connected to the grid at the closest possible connection point that is found using geographic information system (GIS) methods starting with a search radius of 2000 m (Figure 8 (2)). If no points are located inside this circular area, the radius is gradually expanded by 1000 m increments.

(3) If the first two connection options fail due to technical constraints, nodes are connected directly to the main route via separate branch lines using the standard line type (cf. Section 3.3.1). See (Figure 8 (3)).

First, *satellite* LAs run through the grid connection procedure. Second, MV-LV substations are connected to the grid. Then, *aggregated* LAs that are not modelled in detail and therefore directly connected to the bus bar of the HV-MV substation follow, using the line type of largest capacity. Finally, generation units are connected to the grid via branch lines (Table 1).

Grid connection of *satellite* LAs is affected by technical boundary conditions, such as maximum length of and maximum cumulative peak load connected to a branch line. We define a maximum length of branch lines of 2 km [13]. Furthermore, the cumulative peak load of a branch line must not exceed 1 MVA. This is is the maximum power of a typical emergency backup generator used by DSOs to supply branch lines in case of failure within the MV ring [29, 38]. In order to meet the latter condition, the cumulative peak load of *satellite* LAs must be considered. Therefore, we build groups of *satellite* LAs when connecting these to the grid. As

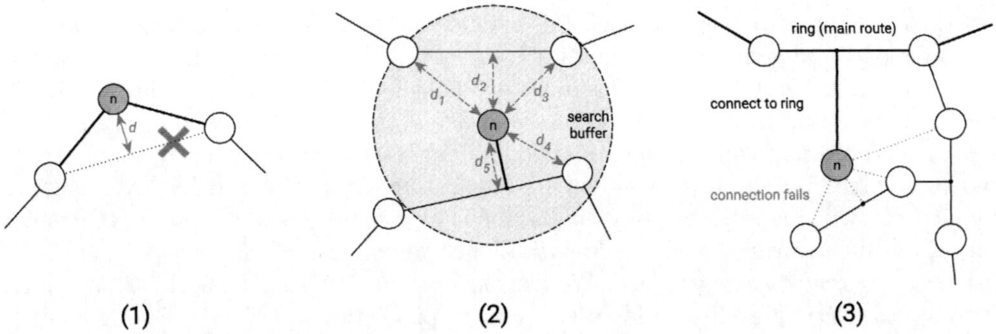

Figure 8: Connect additional nodes: (1) Re-location of line to include node, (2) connect node to closest point within search buffer, (3) fallback method: connect to ring.

grid topology depends on the order of LAs connected to the grid, these are sorted alphabetically to retain a reproducible result. Load areas that cannot be connected to branch lines due to the 1 MVA constraint on cumulative peak load are connected individually to the main route of the MV grid.

Once LAs are connected to the centroid of the area, MV-LV substations inside the LAs are considered, as well as their surrounding supply areas (the LVGDs). The topology of MV grids is refined by connecting these MV-LV substations by the same method applied for connecting *satellite* LAs. In this step, technical constraints such as maximum length and maximum cumulative peak load connected by one branch line are omitted. In addition to the connection of MV-LV substations, overhead lines within LAs are replaced by underground cables. Therefore, we choose a cable type with at least the same current carrying capacity as the replaced overhead line.

As described in Section 3.3.1, a ring's switch disconnector is located on the line where power flow is minimal in closed condition. Due to the extension of the grid, additional loads are connected to the rings. Therefore, the switch disconnectors are subsequently relocated to fulfill this requirement.

The grid connection of *aggregated* LAs takes place at the bus bar of HV-MV substations. These areas of large consumption and dense population are excluded from detailed analyses in the *open_eGo* project, because their grids are assumed to be sufficiently equipped to carry the feed-in by RES. Such an *aggregated* LA is connected by the largest type of cable considered in our paper (see Table A1 in the Appendix).

The last step in extending the base topology is to connect generation units to the grid. These units are added similarly and in compliance with German grid connection codes [18] as described in Section 3.2. The method of grid connection as applied for connecting loads is adapted; technical constraints (maximum branch line length and maximum cumulative capacity) are not considered here.

The grid with extended topology results in a MV grid comprising a number of half-rings which have several branch lines that connect additional loads and generators.

3.4. Grid stability and reinforcement

The base topology of MV grids was tested for current and voltage limit violations (see Section 3.3.1). This does not include branch lines that were added afterwards in order to connect remaining LAs, MV-LV substations and generators. The extension of the base topology by these nodes may threaten grid stability. First, line congestion and overvoltages may exist in these branch lines as those are equipped with the MVGD's standard cable/line type without verifying stable operation of the grid. Second, *satellite* LAs, MV-LV substations and generators add additional load and generation capacities to the grid. These lead to different power flows.

These potential grid stability problems must be identified. Line congestion arises when actual current exceeds rated current of a line/cable. Voltage violations occur when voltage at a node exceeds the tolerable voltage. In general, a DSO must guarantee a voltage range of ±10 % compared to nominal voltage [31]. To allow for separate tests in MV and LV level, this range is usually split into two smaller ranges by DSOs. In MV grids, the connection of RES generators can be approved by ensuring a voltage deviation of at maximum ±2 % relative to voltage levels without RES connected to the grid [18]. We use the open-source tool *PyPSA* [42] for power flow analysis to approve grid stability or identify stability problems.

These problems are subsequently eliminated by grid reinforcement to achieve a computable and stable grid model of the status quo. This can be achieved by using several different methods [30]. A hands-on approach that reflects state-of-the-art procedure of DSOs is described in the *DENA Verteilnetzstudie* [2, 43]. It includes straightforward measures of modifying grid topology and equipment type, both current and voltage-induced grid stability problems can be resolved. Nevertheless, this approach is neither applicable to ring grids with branch lines nor if multiple problems occur at once. Therefore, we use different strategies for grid extension inspired by existing approaches. Firstly, line congestion is resolved by installing a line/cable with appropriate current carrying capacity. Secondly, the resulting grid is tested again for remaining voltage limit violations. These are resolved by replacing affected line segments between the affected location and the HV-MV substation by equipment of the next larger current carrying capacity. Grid segments that are reinforced in case of a voltage violation are on the shortest route between that affected node and the HV-MV substation. If more than one node is affected, these routes are likely to share segments. These segments are reinforced only once. This procedure applies until all voltage limit violations are resolved.

4. Results[1] & discussion

4.1. Grid topology example

In this work, we show that MV distribution grids can be constructed by applying heuristics to solve CVRPs. In this section, we apply all steps described in the previous section to a sample MVGD. We use a rural grid district with a cumulative peak load of 30.1 MVA and a mean load density of 0.16 MVA/km^2. The maximum distance from the HV-MV substation to the LAs' centres is 19 km. Accordingly, it is equipped with 20 kV overhead lines (cf. Section 3.1).

Computing the solution of the savings heuristic in this rural grid district results in five rings with (1) and (4) having suboptimal routes with crossings (see Figure 9). In addition, there is one *aggregated* LA, which is connected to the HV-MV substation.

If we use the *local search* metaheuristics, the solution improves. It has fewer crossings and shorter routes (see Figure 10). The line crossings are removed by the Or-Opt intra-route operator. Due to these alterations, the switch disconnector in ring (1) is relocated to the position of minimal power flow in closed condition (cf. Section 3.3.1). In this example, the inter-route operators do not relocate or exchange nodes between rings.

This network is extended by *satellite* LAs, MV-LV substations and generators (see Figure 11). Using the connection principles of Section 3.3.3, the substations are connected forming an individual grid within each LA. The present MV generators of voltage level 5 (Table 1) are linked to the MV grid while generators of voltage levels 6 and 7 are aggregated inside LVGDs according to their particular locations.

To detect possible line congestion and violations of voltage band, the grid is validated using

[1] Results were found using DINGO v0.1.2 [44]. Data is accessible via `https://doi.org/10.5281/zenodo.795210`

IOP Conf. Series: Journal of Physics: Conf. Series **977** (2018) 012007 doi:10.1088/1742-6596/977/1/012007

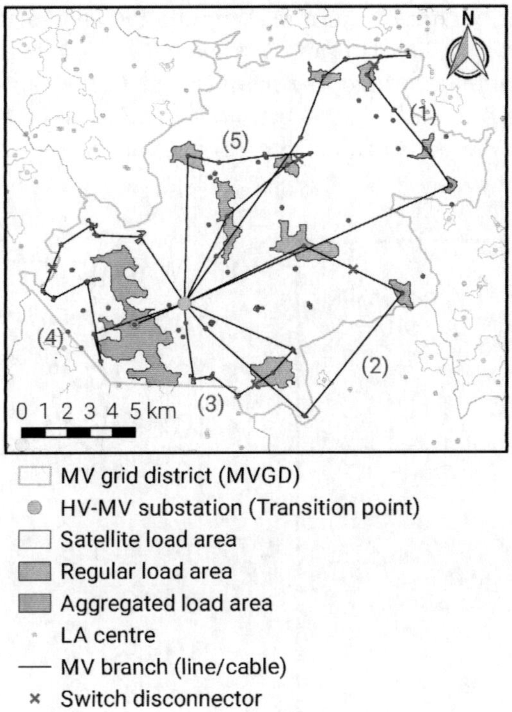

Figure 9: MV rings produced by *savings* heuristic.

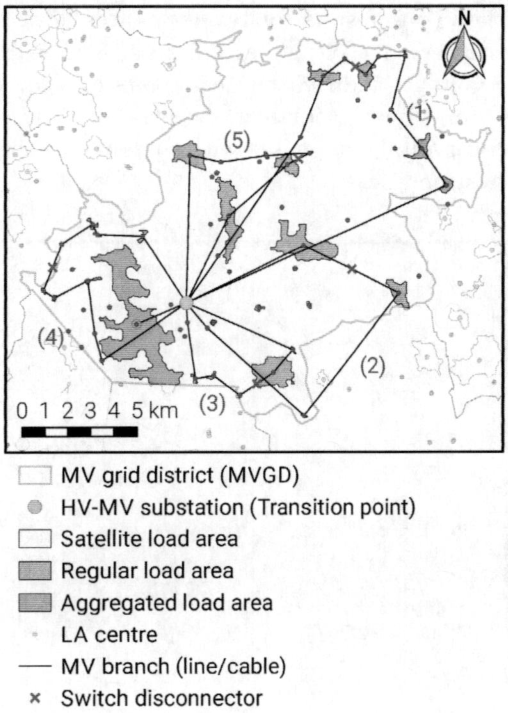

Figure 10: Improved MV rings using *local search* metaheuristics.

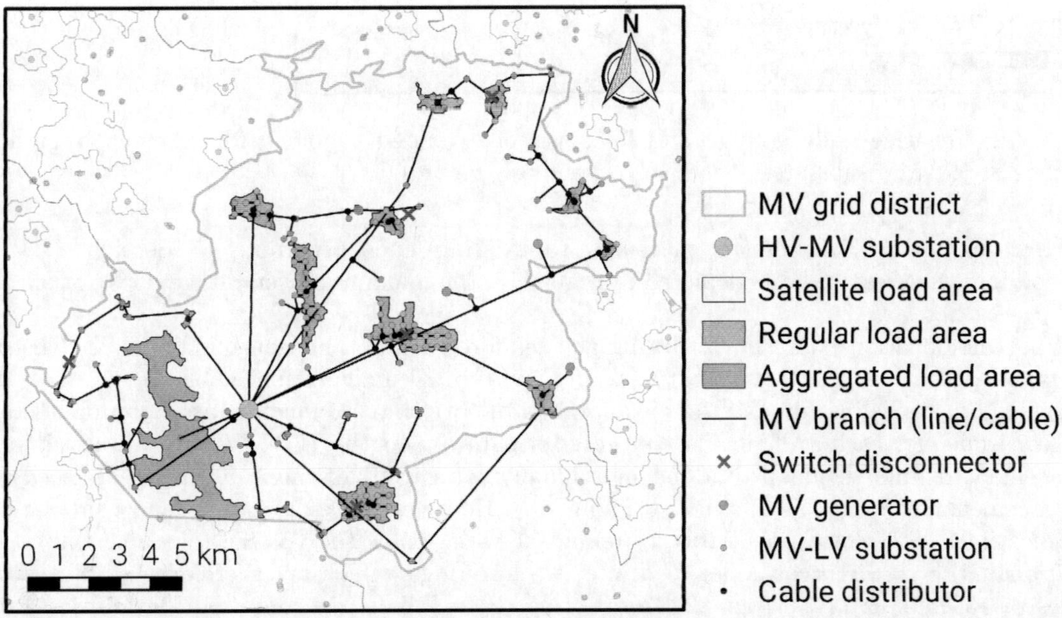

Figure 11: Extended topology with *satellite* load areas, MV-LV substations and generators.

power flow calculations for both the heavy load flow and the reverse power flow scenario. Figure

12 shows the results for reverse power flow. In this scenario, the max. cumulative generation is 2.62 MVA of voltage level 5 and 15.47 MVA of voltage levels 6 and 7 (demand are set to 0). The voltages induced by generators exceed the nominal voltage in rings (1) and (2) by up to 3.7 % and hence violate the tolerable voltage band of ±2 % (cf. Section 3.4). As expected, the overvoltage increases with distance to the HV-MV substation. In contrast, there is no line congestion.

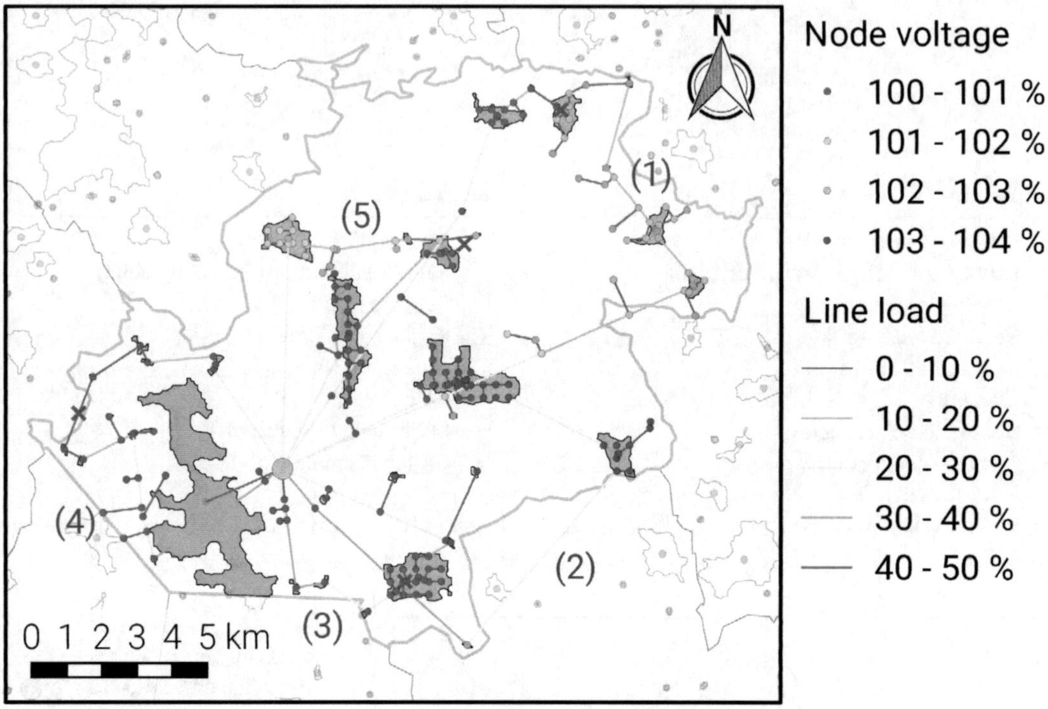

Figure 12: Relative node voltage and line load of extended topology for reverse power flow (voltage at HV-MV substation is set to to nominal voltage).

After applying reinforcement measures to existing lines according to Section 3.4, all overvoltages were resolved. As depicted in Figure 13, the maximum overvoltage does not exceed 2 %.

What modifications were made during the reinforcement? The non-reinforced MV grid is equipped with two types of overhead lines and one type of underground cable (Figure 14), the latter being installed within LAs (cf. Section 3.3.3). Switch disconnectors are in an open state and consequently, each half-ring is connected separately to the HV-MV substation. Hence, replacement of line segments is done individually for each half-ring. There is no need for reinforcements in rings (3), (4) and (5) (Figure 15). However, rings (1) and (2) show substantial modifications. This corresponds to the significant violation of the voltage tolerance caused by additional RES. All critical nodes feed into two half-rings, which are therefore equipped with cables of 150 mm^2 (ring (1)) and 240 mm^2 (ring (2)). Finally, in some segments inside LAs, cables of 240 and 300 mm^2 are used to eliminate overvoltages. This can be traced back to the different types initially used in LAs as described before. Predictably, the line load was decreased due to the line extension.

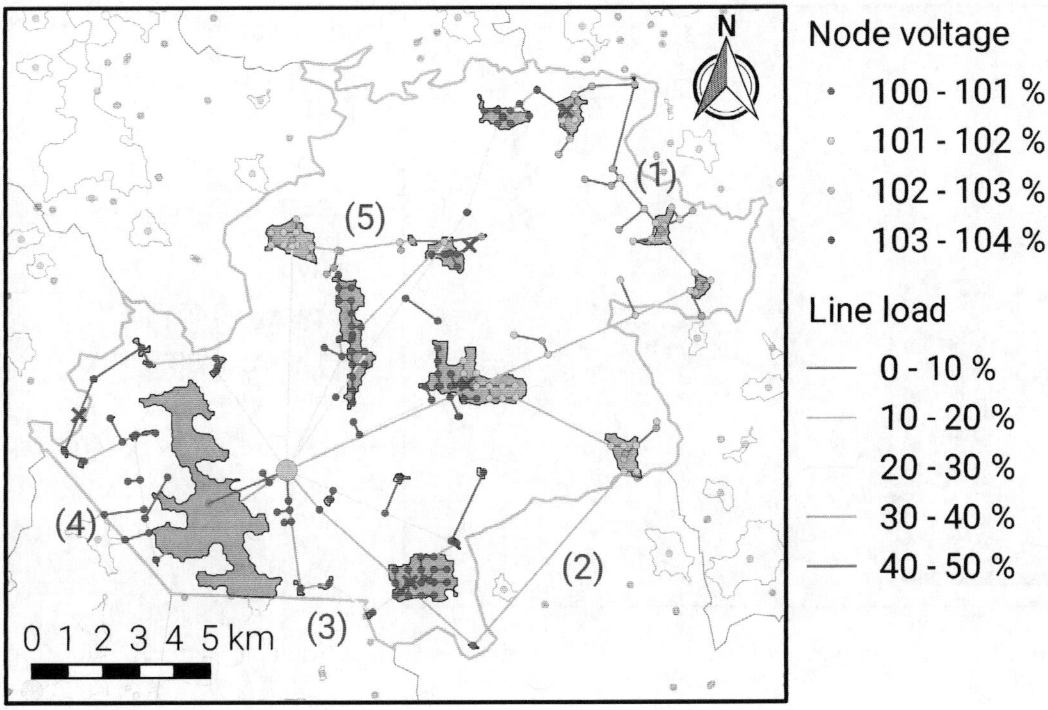

Figure 13: Relative node voltage and line load of extended topology after reinforcement measures.

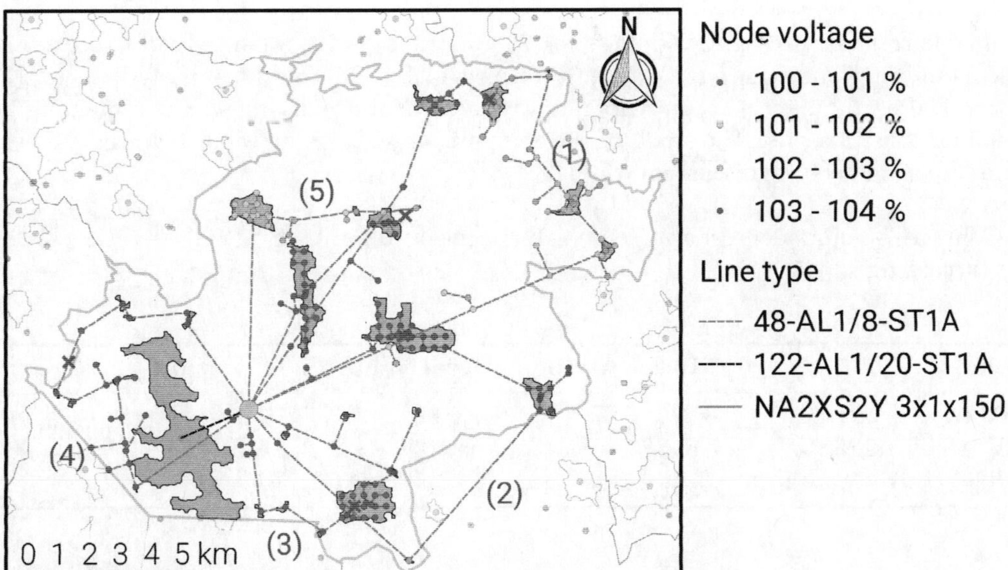

Figure 14: Relative node voltage and line types of extended topology.

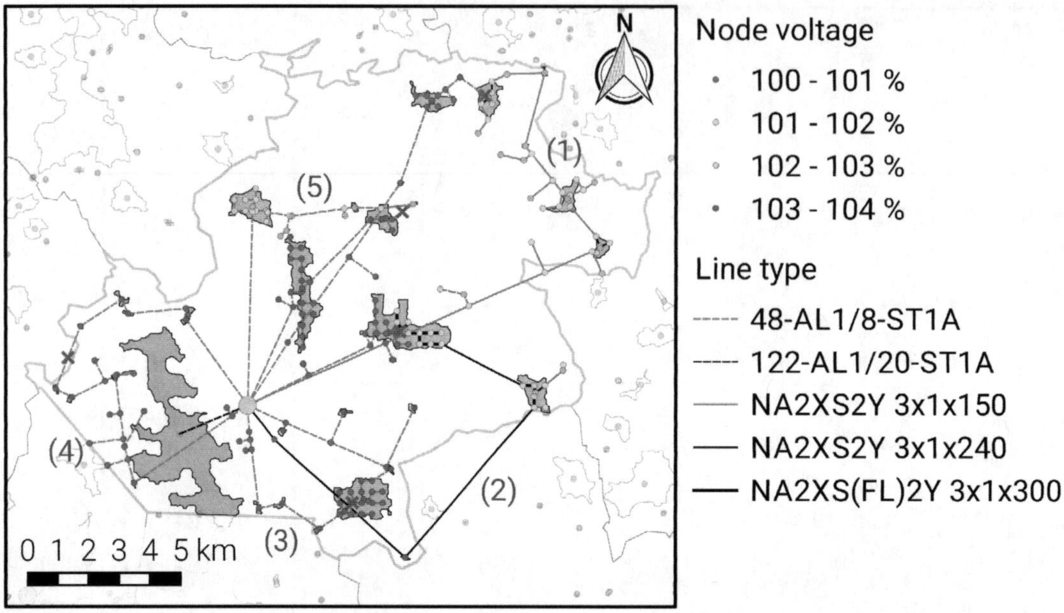

Figure 15: Relative node voltage and line types of extended topology after reinforcement measures.

4.2. Statistical evaluation of MV grids

In order to validate our method, we compared the data of the generated synthetic medium-voltage grids to real network data of Germany. Our data comprises 3608 MVGDs in total. We were unable to generate a grid topology for 239 districts ($\sim 6.6\,\%$) since half of them (119 in total) contain only one or several *aggregated* LAs which are directly connected to the HV-MV busbar. These grid districts are predominantly located in urban areas. The remaining 120 MVGDs fail for other reasons. The result of the comparison is shown in Table 4. In the following the presented parameters are discussed in detail.

Table 4: Comparison of modelled data (some values scaled up for 3,608 MVGDs) with real data (estimates according to [45]).

Equipment	**Model value** m	**Real value** r	**Deviation** $\lvert(r-m)/r\rvert$
HV-MV transformers	8,276	7,500	10.3 %
MV-LV transformers	514,333	560,000	8.2 %
MV cables & overhead lines	495,499 km	507,000 km	2.3 %

HV-MV transformers Because HV-MV substations are usually equipped with at least two transformers to meet the (n-1)-criterion (cf. Section 3.2), our system has 7,728 HV-MV transformers for 3,369 MVGDs that are successfully treated by DINGO. In order to make this number comparable to the reference value of 7,500 found in the literature [45], we assume the same average number of transformers in substations for missing MVGDs. Thus, in total the number of HV-MV transformers for the 3,608 MVGDs adds up to 8,276.

The deviation of 10.3 % can probably be attributed to several reasons. First, the number of HV-MV substations that was determined manually by *Open Street Map* users might be inaccurate. Second, the number of transformers included in a substation is calculated using peak demand and generation capacity of the MVGD. Overestimating these numbers directly affects the number of transformers.

MV-LV transformers Based on an equidistant grid of 360 x 360 m^2 describing the locations of MV-LV substations in LAs, we identified 514,333 MV-LV substations for Germany. We assume that this number also corresponds with the number of MV-LV transformers, since every substation is usually equipped with one MV-LV transformer. The number of identified transformers deviates from values found in the literature by about 8.2 %.

The supply area of 360 x 360 m^2 for each substation is, of course, only an average estimation and is based on the experience of the distribution system operator LVN [29]. However, the mean distance between two substations in rural areas lies within the range of 500 and 2,000 metres and in urban areas between 100 and 400 metres [38]. Hence, the number of substations is overestimated in rural areas and underestimated in urban areas. It should be possible to achieve a more accurate estimation by differentiating between urban and rural areas when modelling the installation of MV-LV substations. Furthermore, we may underestimate the number of transformers by assuming that only one transformer is installed in each substation. Sometimes MV-LV substations are equipped with more than one transformer [46].

MV cables and overhead lines Only 464,180 (of in total 514,333) MV-LV substations are considered in the resulting grid topology. The remaining 50,153 MV-LV substations are located in aggregated load areas (that are of little interest for future studies), which are directly connected to the HV-MV substations by using a virtual substation in the centre of the load area (see Figure 9).

Accordingly, only 477,444 km of cables and overhead lines were laid in the model. This is significantly less than the reference value provided in Table 4. However, there are additional lines supplying the 50,153 substations in *aggregated* LAs which are not included in our model. Since each MV-LV substation supplies an area of 360 x 360 m^2, we assume a line length of 360 metres per substation. This results in a total line length of 495,499 km ($\approx (50,153 \cdot 0.36\,km) + 477,444\,km$). This theoretical total line length exceeds values found in the literature by 2.3 % (cf. Table 4).

4.3. Limitations

The method presented in this paper attempts to mimic characteristics of historic development of *MV* grids. It is designed to reflect local characteristics by using spatially highly disaggregated data. However, some local characteristics which have an impact on a grid's topology are disregarded, such as natural boundaries (e.g. rivers, forests, etc.) as well as transport infrastructure. Furthermore, the large-scale spatial approach cannot consider particularities of solitary cases that often influence grid planning decisions.

The strategy of *local search* produces solutions based on sequential improvements which put restrictions on the solution space [36]. Applying the *local search* strategies in different orders may result in a different optimum. To address this, additional heuristics can be employed, such as large neighbourhood search and guided local search [12]. The first searches in a larger neighbourhood and is therefore less susceptible to be caught in a local optimum. Guided local search controls the *local search* algorithm using penalties in order to stimulate the *local search* to escape local optima.

While calculating the base topology of the MV grid, *regular* LAs define the main route of a ring. The exclusion of smaller loads does not consider if those agglomerate in a certain area.

However, current grid planning principles dictate that agglomerated loads forming a larger load might be treated as part of the main ring. This leads to a tendency for individually connected *satellite* LAs.

The topology of the grid and further characteristics are highly dependent on a set of assumptions. The most influential parameters are: the topology itself (we only account for ring grids), threshold peak load of a *satellite* LA, tolerable length and cumulative branch line peak load, and re-routing distance threshold (cf. Figure 8 (1)).

Another limitation of the identified MV grids is related to the quality of input data. The MVGDs that are used as a fundamental dataset to describe the area supplied by a MV grid are based on a combination of municipal boundaries and Voronoi partition and do not necessarily correspond to the actual shapes of supply areas. Furthermore, the geographical location of RES generators is inaccurate.

It should be noted that the grid topology is reproducible when the same input datasets are used. All methods that use random values are located in the data processing which precedes our tool.

Our method currently does not compute a representation of LV grids. Future publications will cover this topic in detail. Furthermore, sensitivity analyses on crucial parameters as mentioned above will be conducted.

4.4. Comparison with other approaches

The method presented in this paper produces a set of plausible MV grids in Germany. The most distinctive advantages of this dataset are the broad coverage and availability of data. The dataset covers almost all MVGDs in Germany and is publicly accessible through the OEP [17]. Furthermore, due to stringent open-source licensing, the research community can adapt and improve the method. The availability of grid data will allow third parties to evaluate results of studies that use these data. Compared to reference grid data [7, 11, 10], a large range of different grids can be created with the presented methods. Variations of grid topology and equipment properties introduced by Monte-Carlo variations [3] provide a larger variety of grid data. This allows for studies considering various scenarios in DGs returning results of increased robustness. Nevertheless, grid data obtainable from the approach presented in our paper go a step further by considering local characteristics of demand and supply for each grid.

5. Conclusions

To optimise the future electrical energy supply system while considering the interaction of all voltage levels, a holistic approach is required. However, an open dataset for performing this optimisation does not exist. Therefore, a core task of the research project *open_eGo*, funded by the German Federal Ministry for Economic Affairs and Energy, is to close this gap.

The aim of the approach presented in this paper is to model the German status quo distribution system as realistically as possible. While the electricity networks of the high and extra high voltage levels generally run above ground and therefore can be modelled very well in *Open Street Map* using aerial photographs, the modelling of the distribution grid is much more difficult. One reason for this is the fact that the network components at these voltage levels usually are not visible, since underground cables are used and the substations are often integrated into buildings. Furthermore, at these voltage levels a substantially greater number of network components is installed than at higher voltage levels. The challenge is to simplify the modelling of distribution networks, which we address by breaking the problem down, beginning with the medium-voltage networks. A further publication relating to the modelling of low-voltage networks is planned.

All medium-voltage grids are generated for given supply areas using local load and generation data [16]. To enhance our model (results / assumptions) we were able to work with a distribution

IOP Conf. Series: Journal of Physics: Conf. Series **977** (2018) 012007 doi:10.1088/1742-6596/977/1/012007

system operator from southern Germany and use their knowledge about planning principles.

The German electricity networks were designed in the post-war period to cover the energy demand of consumers. Since renewable energy plants have been installed starting in the 1990s, the requirements on the grids' capacity have changed. To account for this development, we choose a two-step approach. Accordingly, in the first step of the modelling process the network topology is exclusively based on the load situation. To create the networks a CVRP is solved, which has a minimum circuit length as target function. In the second step, additional branches with loads as well as renewable energy power plants are connected to the electricity grids resulting from step one. In order to ensure that the generated networks are as realistic and as accurate as possible, typical parameters of medium-voltage components were used. In addition, the stability of these networks was ensured by the use of power flow calculations.

Using this method, the entire medium-voltage level in Germany was modelled. In total 3608 MV networks were created. In terms of the number of transformers and total cable length, the resulting networks show a deviation of $< 10\,\%$ from real network data. It can be concluded that the method presented in this paper allows to create a cost-optimised electrical distribution system for Germany.

Appendix

Table A1: Types of cables and overhead lines considered. Lines and cables of a diameter greater than $300\,\mathrm{mm}^2$ refer to parallel line/cable installation of smaller types.

Type	U_n in kV	$I_{max,th}$ in A	R in Ω/km	L in mH/km	C in μF/km
			Cables		
NA2XS2Y 3x1x185	10	357	0.164	0.38	0.41
NA2XS2Y 3x1x240	10	417	0.125	0.36	0.47
NA2XS2Y 3x1x300	10	466	0.1	0.35	0.495
NA2XS2Y 3x1x400	10	535	0.078	0.34	0.57
NA2XS2Y 3x1x500	10	609	0.061	0.32	0.63
NA2XS2Y 3x1x150	20	319	0.206	0.4011	0.24
NA2XS2Y 3x1x240	20	417	0.13	0.3597	0.304
NA2XS(FL)2Y 3x1x300	20	476	0.1	0.37	0.25
NA2XS(FL)2Y 3x1x400	20	525	0.078	0.36	0.27
NA2XS(FL)2Y 3x1x500	20	598	0.06	0.34	0.3
			Overhead lines		
48-AL1/8-ST1A	10	210	0.35	1.11	0.0104
94-AL1/15-ST1A	10	350	0.33	1.05	0.0112
122-AL1/20-ST1A	10	410	0.31	0.99	0.0115
48-AL1/8-ST1A	20	210	0.37	1.18	0.0098
94-AL1/15-ST1A	20	350	0.35	1.11	0.0104
122-AL1/20-ST1A	20	410	0.34	1.08	0.0106

Nomenclature

CIGRE *Conseil International des Grands Réseaux Électriques*

CVRP capacitated vehicle routing problem

DG distribution grid

IOP Conf. Series: Journal of Physics: Conf. Series **977** (2018) 012007 doi:10.1088/1742-6596/977/1/012007

Table A2: Parameter assumptions applied

Parameter	Value	Unit
Max. branch line load	1	MVA
Max. branch line length	2,000	m
Satellite LA threshold	100	kW
Main route modification threshold	100	m
Grid connection search buffer	2,000	m
Grid connection search buffer increment	1,000	m
10 kV/20 kV load density threshold	1	MVA/km^2
Branch detour factor	1.3	
Max. count of half-rings	20	
$\cos\phi$ (loads)	0.9	
$\cos\phi$ (generators)	1.0	
Nominal operational voltage at HV-MV substation	1.0	p.u.
Max. tolerable voltage deviation (heavy load flow case)	-5	%
Max. tolerable voltage deviation (heavy load flow case, failure)	-10	%
Max. tolerable voltage deviation (reverse power flow case)	+2	%

DS distribution substation

DSO distribution system operator

GIS geographic information system

HV high-voltage

LA load area

LF load factor

LV low-voltage

LVGD low-voltage grid district

LVN *LEW Verteilnetz GmbH*

MV medium-voltage

MVGD medium-voltage grid district

NEP *Netzentwicklungsplan* (Network Development Plan)

OEP *OpenEnergy Platform*

OSM *Open Street Map*

RES renewable energy sources

WEC wind energy converter

WPA wind potential area

References

[1] Rippel K M, Wiede T, Meinecke M and König R 2017 Netzentwicklungsplan Strom 2030, Version 2017 – Erster Entwurf der Übertragungsnetzbetreiber Available online: https://www.netzentwicklungsplan.de/de/netzentwicklungsplaene/netzentwicklungsplaene-2030

[2] Agricola A C *et al.* 2012 Ausbau- und Innovationsbedarf der Stromverteilnetze in Deutschland bis 2030 (dena-Verteilnetzstudie): Endbericht Tech. rep. German Energy Agency (dena) Berlin available online: https://shop.dena.de/fileadmin/denashop/media/Downloads_Dateien/esd/9100_dena-Verteilnetzstudie_Abschlussbericht.pdf

[3] Büchner J, Katzfey J, Flörcken O and Moser A 2014 Moderne Verteilernetze für Deutschland (Verteilernetzstudie) Tech. rep. E-Bridge, IAEW, OFFIS Available online: http://www.bmwi.de/Redaktion/DE/Publikationen/Studien/verteilernetzstudie.html

[4] IKS OvGU, Next Energy, RLI, ZNES 2015 Netzebenen-übergreifendes Planungsinstrument - zur Bestimmung des optimalen Netz- und Speicherausbaus in Deutschland - integriert in einer OpenEnergy-Plattform Available online: https://openegoproject.wordpress.com

[5] Müller U P, Wienholt L, Kleinhans D, Cußmann I, Bunke W D, Pleßmann G and Wendiggensen J 2017 The eGo grid model: An open source approach towards a model of German high and extra-high voltage power grids *SciGRID International Conference on Power Grid Modelling* (submitted)

[6] Medjroubi W, Müller U P, Scharf M, Matke C and Kleinhans D 2017 Open data in power grid modelling: New approaches towards transparent grid models *Energy Reports* **3** 14–21 URL http://www.sciencedirect.com/science/article/pii/S2352484716300877

[7] Pilo F *et al.* 2014 *Benchmark Systems for Network Integration of Renewable and Distributed Energy Resources* (Conseil international des grands réseaux électriques. Comité d'études C6) ISBN 9782858732708

[8] Bundesministerium der Justiz und für Verbraucherschutz 2005 Verordnung über die Entgelte für den Zugang zu Elektrizitätsversorgungsnetzen (Stromnetzentgeltverordnung – StromNEV) Available online: https://www.gesetze-im-internet.de/stromnev/index.html

[9] Bundesministerium der Justiz und für Verbraucherschutz 2005 Verordnung über den Zugang zu Elektrizitätsversorgungsnetzen (Stromnetzzugangsverordnung – StromNZV) Available online: https://www.gesetze-im-internet.de/stromnzv/index.html

[10] Rui H, Arnold M and Wellssow W H 2012 Synthetic Medium Voltage Grids for the Assessment of Smart Grid Techniques *2012 3rd IEEE PES Innovative Smart Grid Technologies Europe (ISGT Europe)* (Berlin: IEEE) pp 1–8

[11] Schaller F, Karstädt F, Warweg O and Bretschneider P 2011 Modellierung realitätsnaher zukünftiger Referenznetze im Verteilnetzsektor zur Überprüfung der Elektroenergiequalität *ETG-Fachbericht-Internationaler ETG-Kongress 2011* (VDE VERLAG GmbH)

[12] Tao X, Haubrich H J and Maurer C 2006 Automatisierte Grundsatzplanung für Mittelspannungsnetze *ET. Energiewirtschaftliche Tagesfragen* **56** 8–11

[13] Consentec GmbH 2006 Untersuchung der Voraussetzungen und möglicher Anwendung analytischer Kostenmodelle in der deutschen Energiewirtschaft Tech. rep. Consentec GmbH available online: https://www.bundesnetzagentur.de/SharedDocs/Downloads/DE/Sachgebiete/Energie/Unternehmen_Institutionen/Netzentgelte/Anreizregulierung/GA_AnalytischeKostenmodelle.pdf

[14] Strunz K *et al.* 2014 *Planning and Optimization Methods for Active Distribution Systems* (Conseil international des grands réseaux électriques. Comité d'études C6)

[15] Harnisch S, Steffens P, Thies H, Monscheidt J, Münch L, Böse C and Gemsjäger B 2016 Planungs- und Betriebsgrundsätze für ländliche Verteilungsnetze - Leitfaden zur Ausrichtung der Netze an ihren zukünftigen Anforderungen Tech. rep. Bergische Universität Wuppertal Available online: http://elpub.bib.uni-wuppertal.de/servlets/DerivateServlet/Derivate-5651/eaw0216.pdf

[16] Hülk L, Wienholt L, Cußmann I, Müller U P, Matke C and Kötter E 2017 Allocation of annual electricity consumption and power generation capacities across multiple voltage levels in a high spatial resolution *International Journal of Sustainable Energy Planning and Management* (submitted)

[17] IKS OvGU, Next Energy, RLI, ZNES 2017 OpenEnergy Platform Available online: http://oep.iks.cs.ovgu.de/ (launches in April 2017)

[18] Bundesverband der Energie- und Wasserwirtschaft eV (BDEW) 2008 Technische Richtlinie Erzeugungsanlagen am Mittelspannungsnetz - Richtlinie für Anschluss und Parallelbetrieb von Erzeugungsanlagen am Mittelspannungsnetz

[19] Bundesnetzagentur 2015 List of power plants Available online: http://www.bundesnetzagentur.de/cln_1911/DE/Sachgebiete/ElektrizitaetundGas/Unternehmen_Institutionen/Versorgungssicherheit/Erzeugungskapazitaeten/Kraftwerksliste/kraftwerksliste-node.html

[20] Deutsche Gesellschaft für Sonnenenergie eV 2014 Die EEG-Anlagen der Region Bundesrepublik Deutschland (EnergyMap) Available online: http://www.energymap.info/energieregionen/DE/105.html

[21] Gerbaulet C and Kunz F 2016 Data provided by Open Power System Data - Data Package Conventional power plants, version 2016-10-27. Primary data from BNetzA Kraftwerksliste, Umweltbundesamt Datenbank Kraftwerke in Deutschland Available online: http://open-power-system-data.org/

[22] Bunke W D 2016 Data provided by Open Power System Data - Data Package Renewable power plants, early version 2016-02-10. Primary data from BNetzA, BNetzA_PV, TransnetBW, TenneT, Amprion, 50Hertz, Netztransparenz.de, Postleitzahlen Deutschland, Energinet.dk, Energistyrelsen, GeoNames, French Ministery of the Environment, Energy and the Sea, OpenDataSoft, Urzad Regulacji Energetyki (URE) Available online: http://open-power-system-data.org/

IOP Conf. Series: Journal of Physics: Conf. Series **977** (2018) 012007 doi:10.1088/1742-6596/977/1/012007

[23] Degel M, Christ M, Becker L, Grünert J and Wingenbach C 2016 VerNetzen – Sozial-ökologische und technisch-ökonomische Modellierung von Entwicklungspfaden der Energiewende Tech. rep. Europa-Universität Flensburg Available online: http://www.uni-flensburg.de/fileadmin/content/abteilungen/industrial/dokumente/downloads/veroeffentlichungen/forschungsergebnisse/vernetzen-2016-endbericht-online.pdf
[24] Lütkehus I, Salecker H and Adlunger K 2013 Potenzial der Windenergie an Land Tech. rep. German Environment Agency (UBA) Available online https://www.umweltbundesamt.de/sites/default/files/medien/378/publikationen/potenzial_der_windenergie.pdf
[25] Kerber G 2011 *Aufnahmefähigkeit von Niederspannungsverteilnetzen für die Einspeisung aus Photovoltaikkleinanlagen* Ph.D. thesis Technische Universität München
[26] Scheffler J 2002 *Bestimmung der maximal zulässigen Netzanschlussleistung photovoltaischer Energiewandlungsanlagen in Wohnsiedlungsgebieten* Ph.D. thesis TU Chemnitz
[27] Mohrmann M, Reese C, Hofmann L and Schmiesing J 2012 Untersuchung von Niederspannungsverteilnetzen anhand von synthetischen Netzstrukturen *VDE Kongress 2012 Smart Grid: Intelligente Energieversorgung der Zukunft* (VDE)
[28] Heuck K, Dettmann K D and Schulz D 2010 *Elektrische Energieversorgung* 8th ed (Wiesbaden: Vieweg+Teubner)
[29] LEW Verteilnetz GmbH 2016 *open_eGo* workshop: Method and data basis for generation of medium- and low-voltage grids (private communication)
[30] Resch M, Bühler J, Klausen M and Sumper A 2017 Impact of Operation Strategies of Large Scale Battery Systems on Distribution Grid Planning in Germany *Renewable and Sustainable Energy Reviews* **74** 1042–1063 URL http://www.sciencedirect.com/science/article/pii/S1364032117302976
[31] German Institute for Standardisation (DIN) 2011 Voltage characteristics of electricity supplied by public distribution networks; German version EN 50160:2010 + Cor.: 2010
[32] Verband der Elektrotechnik Elektronik Informationstechnik eV (VDE) 2011 VDE-AR-N 4105: Erzeugungsanlagen am Niederspannungsnetz – Technische Mindestanforderungen für Anschluss und Parallelbetrieb von Erzeugungsanlagen am Niederspannungsnetz
[33] Flosdorff R and Hilgarth G 2005 *Elektrische Energieverteilung* 9th ed (Wiesbaden: Vieweg+Teubner)
[34] Wenger W 2010 *Multikriterielle Tourenplanung* Gabler Research : Produktion und Logistik (Gabler Verlag)
[35] Stürmer J 2001 *Instandhaltungs- und Erneuerungsstrategien in Verteilungsnetzen* Ph.D. thesis Universität Dortmund
[36] Bräysy O and Gendreau M 2005 Vehicle Routing Problem with Time Windows, Part I: Route Construction and Local Search Algorithms *Transportation Science* **39** 104–118 URL http://dx.doi.org/10.1287/trsc.1030.0056
[37] Clarke G and Wright J W 1964 Scheduling of Vehicles from a Central Depot to a Number of Delivery Points *Operations Research* **12** 568–581
[38] Forschungsgemeinschaft für Elektrische Anlagen und Stromwirtschaft e V 2008 Technischer Bericht 302: Ein Werkzeug zur Optimierung der Störrungsbeseitigung für Planung und Betrieb von Mittelspannungsnetzen Tech. rep. Forschungsgemeinschaft für Elektrische Anlagen und Stromwirtschaft e. V.
[39] Engels K 2000 *Probabilistische Bewertung der Spannungsqualität in Verteilungsnetzen* Aachener Beiträge zur Energieversorgung (Klinkenberg-Verlag)
[40] Sakulin M and Hipp W 2004 Netzaspekte von dezentralen Erzeugungseinheiten, Studie im Auftrag der E-Control GmbH Tech. rep. TU Graz Available online: https://www.e-control.at/documents/20903/-/-/1109e85c-63ca-4add-8c7e-a6376e309acd
[41] Boomgaarden C 2007 *Dynamische Tourenplanung und -steuerung* Ph.D. thesis Universität Passau
[42] Brown T, Hörsch J and Schlachtberger D 2017 PyPSA: Python for Power System Analysis *ArXiv e-prints* (*Preprint* 1707.09913)
[43] Ackermann T, Untsch S, Koch M and Rothfuchs H 2014 Verteilnetzstudie Rheinland-Pfalz Tech. rep. energynautics GmbH, Öko-Institut e.V., Bird & Bird LLP
[44] Amme J and Pleßmann G 2017 Distribution network generator (dingo) URL https://doi.org/10.5281/zenodo.844045
[45] Brunekreeft G, Luhmann T, Menz T, Müller S and Recknagel P 2015 *Regulatory Pathways For Smart Grid Development in China* 1st ed (Springer Vieweg)
[46] Bühler J 2013 *Instandhaltungs- und Erneuerungsoptimierung von städtischen Mittelspannungsnetzen* Ph.D. thesis TU Darmstadt

IOP Conf. Series: Journal of Physics: Conf. Series **977** (2018) 012008 doi:10.1088/1742-6596/977/1/012008

Evaluation of load flow and grid expansion in a unit-commitment and expansion optimization model

SciGRID International Conference on Power Grid Modelling

Charlotte Senkpiel*[1], Wolfgang Biener[1], Shivenes Shammugam[1] and Sven Längle[1]
[1]Fraunhofer Institute for Solar Energy Systems ISE, Freiburg, Germany
*charlotte.senkpiel@ise.fraunhofer.de

Abstract. Energy system models serve as a basis for long term system planning. Joint optimization of electricity generating technologies, storage systems and the electricity grid leads to lower total system cost compared to an approach in which the grid expansion follows a given technology portfolio and their distribution. Modelers often face the problem of finding a good tradeoff between computational time and the level of detail that can be modeled. This paper analyses the differences between a transport model and a DC load flow model to evaluate the validity of using a simple but faster transport model within the system optimization model in terms of system reliability. The main findings in this paper are that a higher regional resolution of a system leads to better results compared to an approach in which regions are clustered as more overloads can be detected. An aggregation of lines between two model regions compared to a line sharp representation has little influence on grid expansion within a system optimizer. In a DC load flow model overloads can be detected in a line sharp case, which is therefore preferred. Overall the regions that need to reinforce the grid are identified within the system optimizer. Finally the paper recommends the usage of a load-flow model to test the validity of the model results.

1. Introduction

In the Paris agreement, 195 countries adopted the first ever legally binding global climate agreement. The goal is to keep the global average temperature well below 2 degree Celsius (European Commission, 2016). The latest official document of the German government is the "Klimaschutzplan 2050" [1], which confirms the targets of the "Energiekonzept" of 2010 [2] and sets additional CO2-reduction targets for each sector. To reduce the CO2-emissions and to increase the share of renewable energies, energy system models can add value in developing a strategy for the system planning, in providing possible pathways. The models are able to take the costs of the technologies, the different local potentials as well as the existing infrastructure into account. The energy system model ENTIGRIS for example integrates the unit-commitment, transmission and expansion planning of generating technologies, storage systems as well as grid expansion into a cost minimal optimization. Disregarding the grid would lead to different transformation pathways, compared to an approach where the grid expansion is optimized endogenously with the technology and storage systems. This will additionally lead to higher system costs [3]. The challenge for the model set up is to find a good trade-off between computing time and the level of detail. The level of detail is mainly characterized by number of regions, in which the system is clustered, number of optimized time steps and years that are calculated, the aggregation of single power plants as well as the detailing of the grid. This paper therefore analyses the performance of the energy system optimizer ENTIGRIS in terms of grid representation using a comparison to a DC load flow model. Within this paper the use case of Baden-Württemberg was chosen to analyse the effect of grid representation within the highly meshed overlaying grid (see Figure 4). Different levels of detail concerning the number of regions as well as an aggregation of grid lines compared to a line sharp approach will be discussed, regarding the root mean squared error (rmse) of line usage, the rmse of power flows as well as the grid expansion and cases of overload.

2. State of the art

The approach of combining long-term system optimization with grid expansion was evaluated in studies like [3] or [4]. Positive benefits like lower system costs and higher system stability related to a more equally distribution of the generation technologies and therefore lower congestion can be attained in joint system planning. There are different approaches to represent the load flow. [3, 4] point out the relation between the model fidelity and the computational time and efforts. They defined three categories of representing the load flow, which differ in the degree of accuracy:

The **AC optimal power flow** represents the real and reactive power flows abiding physical laws. These are represented in non-linear functions. It is the most challenging method but also the one with the best fidelity. The problem can be formulated as a mixed integer non-linear problem (MINLP). The **decoupled load flow (DC)** model is used as a typical market and planning application. It has two parts, the real power flow (directly proportional to angle differences) and reactive power flow (directly proportional to bus voltage differences). For planning a system with minimal system cost the problem is focused on real power flow represented as a linear problem. Calculating the expansion of a grid for a meshed network like the German overlaying grid Kirchhoff's voltage law turns the optimization problem into a MINLP. Using binary variables instead of integer variables the problem can be formulated as a mixed integer linear problem (MILP). But these problems also require high computational times and cause inaccuracies which violate the voltage criteria. These violations cannot be detected using this methodology. The third approach is the **transshipment model** or simple transport model, which has the advantage that computational times are better. The disadvantage is that the model fidelity decreases due to neglecting the voltage variables and the relationship of real power transfers with bus angle differences and line impedances. This can result in an unfeasible system in reality. Energy system models like Remix and Power ACE use a transport model to represent the high voltage DC lines for Europe (each country represented by one node) [5] [6]. A benefit of this approach is that the computational time is applicable also for large scale optimization problems. Another approach is the **one node model** or "Copper-plate" approach, which neglects the grid in the energy system. The assumption is that grid congestion is not an issue [6].

[3] suggest that the expansion results of the transshipment models need to be checked with a DC or AC model, to ensure the security of the network. [7] present(s) a methodology to couple an electricity market model with a power flow model in order to jointly optimize both power generation and transmission grid infrastructures under flow-based market coupling using an iterative approach based on power transfer distribution factors (PTDFs)." Having the non-linear dependence of the impedances [7] suggests an update of the matrix after each grid reinforcement. In contrast to this paper [7] uses the partial equilibrium model DIMENSION, which takes the shares and the mix of RE technologies as an input. For the European system the approach converges after 5 iteration steps. For the ENTIGRIS Model optimizing unit-commitment of all power plants, storage, transport between the regions and power plant and storage as well as grid expansion at the same time in a non-myopic set up, an iterative approach would lead to extreme high computational times (e.g. using around 4000 hours and 50 nodes for Germany result in one optimization run of several weeks). Therefore the authors chose to use a transshipment model within the ENTIGRIS optimization model and validate the results based on a DC load flow approach presented by [12] (described in 3.3.), which is similar to the approach of [7] as it

IOP Conf. Series: Journal of Physics: Conf. Series **977** (2018) 012008 doi:10.1088/1742-6596/977/1/012008

updates the impedances after each iteration but formulates the problem with voltage angles, whereas [7] uses PTDF matrices

3. Methodology

According to [3] for using an AC approach high computational effort is needed and also data on control devices is necessary. As the target is to optimize the long term development of an energy system like Germany in the highest resolution possible, the power generators in a model region are aggregated. Therefore there is a lack of data on generator capability curves and the AC optimal flow is hence not applicable. The preferable model framework in the context of long-term system planning, takes high regional resolution and high number of time steps into account. As this level of detail will increase the computational time, a DC load flow approach within the optimization model would further increase the computational effort considerably. Hence, the authors choose to combine two models. One model being the ENTIGRIS optimization model (see chapter 3.2.) which optimizes the unit-commitment of renewables, short and long-term storage technologies and the im- and exports between regions as well as the expansion of all available technologies (including grid expansion). The resulting nodal transmission balance and the new grid infrastructure are then transferred into a load flow and grid expansion model based on [12], referred to as DC-load flow model. The DC model provides expansion or upgrade suggestions in the case of overloads, if a potential to do so exists. The DC load flow model therefore ensures that the ENTIGRIS optimization ensures a stable system (in the cases without overloads). Additionally the DC load flow model serves as a benchmark for the calculated load flows and the grid expansion within the ENTIGRIS model as the DC model is more accurately in physical load flow than the simple transport model. Different approaches of grid representation are tested in order to obtain good results. In doing so, following research questions are defined.

3.1. Research Questions

The following research questions are defined within this paper:

- What is the effect of the level of regional resolution (using the highest regional resolution possible and clustering different regions to one model region)?
- What is the effect of a detailed grid representation (representation of each line between regions) compared to a simplified grid (aggregation of lines between regions) within a system optimizer?
- What is the effect of the representation of the load flow by the reactance or the line length within the optimization model?
- How accurate is the performance of ENTIGRIS grid expansion validated by DC load flow?

3.2. Description of the ENTIGRIS optimization model

The objective of ENTIGRIS is to minimize the total system costs of the electric power sector by considering the unit-commitment of power plants and storage systems under the restriction of the grid infrastructure and the power plant expansion (including renewable technologies, storage systems and the electric transmission grid). The description of the main methodology can be found in [8], [9] and [10]. The model considers the electricity sector and the following technologies: Wind power plants, photovoltaics, biogas, CCGT, OCGT, lignite and hard coal, nuclear, hydropower as well as storage technologies (pump storage, batteries). The renewable energy potential is available in NUTS3 level. The 220 and 380 kV voltage grid is integrated in the model. Costs for load change as well as load change constraints are formulated linearly in the model. The power plants of similar types are aggregated within one model region (assuming an average efficiency in dependence of the age of the power plants). The main model scope is displayed in (Figure 1). The annuity costs are broken down to the annuity that occurs for each modelled time step. The annuity payments that have to be undertaken after the last optimization year for an investment that was undertaken within the optimization time are cut off. This enables the model to have the correct ratio of fix to variable costs, even if not all hours of the year are considered. The sum of the annuity costs per optimization year are discounted by a pre-calculated discount factor, including the life time per technology and taking into account that only selected years are optimized (e.g. 2020, 2030, 2040, 2050). This enables the model to minimize the cost for all chosen time steps and optimization years simultaneously.

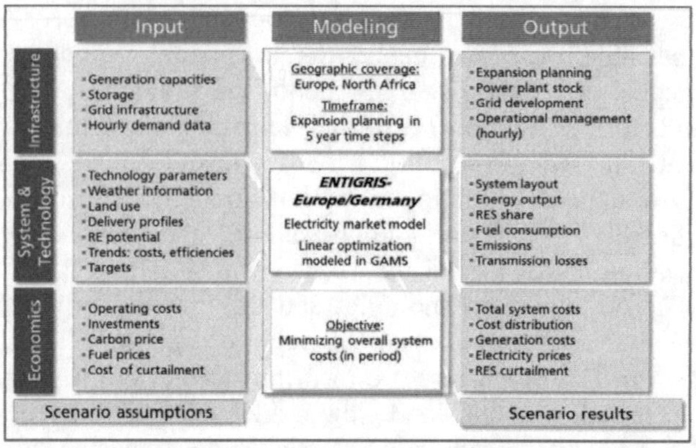

Figure 1: Scheme of ENTIGRIS expansion and unit-commitment optimization model

The balance equation ensures that the sum of generation, storage charge discharge, import and export energy in each node (model region) has to be congruent with the electrical demand, which follows the theory of nodal pricing, as the unit-commitment is restricted by the given grid infrastructure.
The electricity balance equation is used in different forms in this paper. A reference case is calculated for both regional variations (compare 3.4.1.) using the ENTIGRIS optimization for the whole infrastructure (generation, storage, grid) while the grid expansion being switched off for 2020 to ensure a standard case for comparison between the DC load flow and ENTIGRIS. Different variations (explained in 3.4.) are then optimized related to power-flows and grid expansion. Therefore a fixed generation time series is used from a reference scenario, to be able to analyze the effects of the variations directly, without a changing optimal generation portfolio.
To restrict the load flow in the model, a constraint that limits the possible transfer to the given or optimized line capacity including a security factor of 0.7, representing the n-1 criteria, is used. The grid expansion is represented as a mixed integer linear problem (MILP). The following options are integrated: A 220 kV line can be upgraded to a 380kV line and new 380 kV lines can be installed. Different investment costs (€/km) are associated with the two options. The potential is given by a maximum number of systems of two that can be installed on a power line corridor (link between two regions with a given length). If lines are aggregated the potential sums up accordingly. The construction of new corridors is not included in the calculations for this paper. The results presented in this paper display the minimal system cost under the constraint to fulfill a renewable energy share of 20% in 2020, 40% in 2030, 60% in 2040 and 80% in 2050 of the electricity demand in sum over all modeled regions. The system optimization results in an increase of the wind capacities from 9.6 to 18 GW whereas the PV capacities are decreasing from 12.5 to 7 GW between 2020 and 2050. Only those regions that are displayed in Figure 4 are optimized and were treated as an island system for this analysis.

3.3. *The DC load-flow model.*
The DC load-flow methodology is chosen to calculate grid states since there is not enough information of the grid available for a more accurate AC load-flow calculation. Assumptions made for the DC load flow model are that voltage magnitudes at each node are equal to the nominal voltage and resulting power-flow and only depend on line reactances [11]. [12] shows that this problem may also be formulated as linear optimization problem which allows to implement grid reinforcement options. The optimization problem proposed in [12] is applied for this paper. In the case of overload a line is

IOP Conf. Series: Journal of Physics: Conf. Series **977** (2018) 012008 doi:10.1088/1742-6596/977/1/012008

upgraded or expanded, if there is a remaining potential. The decision if a line is upgraded or a new line is build is based on minimal cost, ensuring that the capacity needed is provided. Having a case of overload at two different grid corridors both corridors will be reinforced. If no potential exists, in the case of overload no possible optimization solution can be provided. In order to hold linearity, changes in line properties due to line upgrades are not integrated in the optimization problem. If line reinforcement was chosen by the optimizer the grid parameter are recalculated according to the new structure and a iteration is performed.

3.4. Description of the reference cases

Three criteria with different variants are identified in order to answer the research questions. The target is to evaluate the performance of the ENTIGRIS model in terms of valid grid expansion. The grid in both models is represented according to the ELMOD-DE model [13].

3.4.1. Variation 1: Clustering of area. The highest regional resolution for this paper, in which the necessary data for power system optimization are fully available, is NUTS 3 level (small administrative units according to the nomenclature of territorial units for statistics [14]. The number of regions has a direct impact on the calculation time of the model. Two cases are selected for this paper. Both have the regional coverage of Baden-Württemberg and its neighboring regions (compare Figure 4). Case one, referred to as BaWü-Nuts3_c, is a slightly simplified presentation of the regional resolution, in which regions are clustered to neighboring regions that have no own node to the ultra-high voltage transmission grid. 63 model regions represent that case. The second case, referred to as BaWü-HVTSO represents a cluster of the regions based on the high voltage transmission system operators, resulting in nine regions. The grid needs to be clustered for this case by using a cross border approach, considering only the lines that cross two regions. The power lines within the clustered region are neglected.

3.4.2. Variation 2: Level of detail of the grid representation. Two different levels of grid detail are implemented in both models: On the one hand, all lines that stretch from one region to another are aggregated. The line length (which is important for calculating the costs of grid expansion) is aggregated by using the arithmetic mean. This approach is defined as the aggregated case, only representing one virtual corridor between two regions for each voltage level), resulting in 21 corridors for HVTSO and 99 corridors for NUTS3_c. On the other hand, the line-sharp case represents aggregates only the lines that have the same voltage level and line length, resulting in 48 lines for HVTSO and 127 lines for NUTS3_c. For all cases just one node (represented by demand and supply) per region is assumed.

3.4.3. Variation 3: Representation of the load flow in the optimization model. A variation of the side constraint of transmission cost is conducted. In one case line length and in the second the line reactance is used for calculating the transmission cost in order to evaluate the load flows and grid expansion in the ENTIGRIS model. Both parameters will be normalized to its maximum value in the reference system and then multiplied with variable costs of 0.1 €/kWh and with the transported electricity (kWh).

3.4.4. Criteria for the analysis. The focus of the analysis is twofold. Firstly, the **load-flow** of both scenarios according to respective models is analyzed. In doing so, the degree of usage of each power line is determined. The root-mean-square-error (RMSE) of the degree of usage is then calculated compared to the DC load flow model. This is conducted for the optimization year of 2020, where no expansion takes place. Critical power lines and the frequency of occurrence are determined by identifying the load-flow in power lines that exceeds the permissible capacity in the DC load flow model. Secondly, the **expansion** of the power grid lines for the optimization years of 2030, 2040 and 2050 is analyzed for both models. The amount and capacity of upgraded lines as well as the deployment of new lines are compared. If the DC load flow model has no expansion suggestions, the system stability is ensured. However, if any expansion is suggested, then the grid capacity of at least one line in the simple transport model is insufficient.

4. Results

Figure 2 shows the RMSE of the power flows between the described scenarios. It can be observed that the aggregated scenarios for both regional resolutions have lower outliers than the detailed case. This can be explained by the existence of only one corridor between two regions in the aggregated case. Therefore in the detailed case the deviations can be very high due to the fact that other lines from different corridors are used within the optimization model than in the DC load flow model. In the case of a higher regional resolution the overall deviation is lower than that in the HVTSO case, whereas the median is around 200 MW in both the NUTS3 and the HVTSO detailed case. The load flow of the optimization model and the DC load flow deviate in all cases in the range of 30 to 1200 MW.

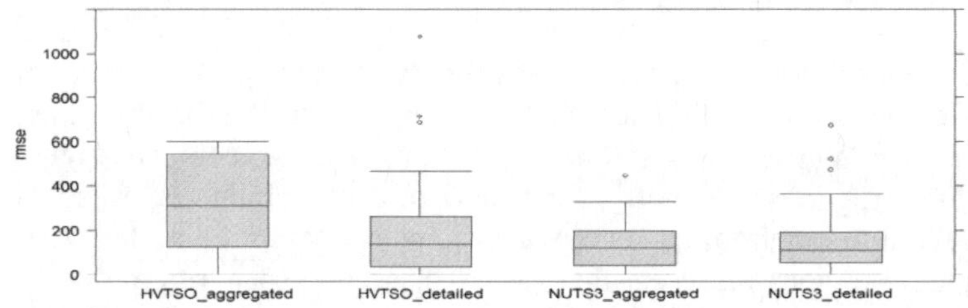

Figure 2: RMSE of power flows per grid line (detailed case) and corridor (aggregated) in MW

Figure 3 shows the degree of usage of each power line in all scenarios. The degree of usage in the detailed scenarios is more scattered than their respective counterparts. This is due to the higher possibilities of lines in the detailed case. The highest outlier is 70% in the HVTSO detailed scenario. The mean deviation in all scenarios is around 10%. This means that an aggregation of lines between regions is comparable to the detailed representations of the power lines. A parameter variation is conducted by varying the variable costs and the line impedance in order to further reduce the deviations between both models. However, no significant trends are observed as the load flows are only minimally affected.

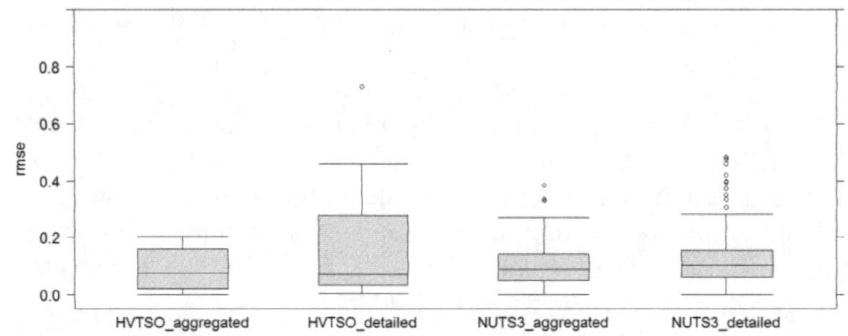

Figure 3: RMSE between DC load flow and transport model of line usage for the four scenarios'

The second part of the analysis concentrates on the validity of the ENTIGRIS results, represented by the cases of overload that occur in the DC load flow model, using the calculated grid by ENTIGRIS. In the HVTSO case the aggregated modelling result in one occurrence of overload in 2020. A reinforcement on this corridor result in a valid system. The expansions by ENTIGRIS in the following optimization years result in a valid system, because the critical line was expanded by ENTIGRIS. In contrast to that the line sharp case (see Figure 4) result in an overload on the same corridor in 2020 but the corresponding rather short line cannot be expended due to potential restrictions. The expansions that were calculated by ENTIGRIS in 2030 for that line is not valid for the DC load-flow; again cases

IOP Conf. Series: Journal of Physics: Conf. Series **977** (2018) 012008 doi:10.1088/1742-6596/977/1/012008

of overload occur in 2030, 2040 and 2050. In 2040 an additional corridor faces the case of overload. The bottleneck effect of a small line within a meshed network is better represented using the line sharp model and cannot be detected using aggregated lines, in the DC load-flow.

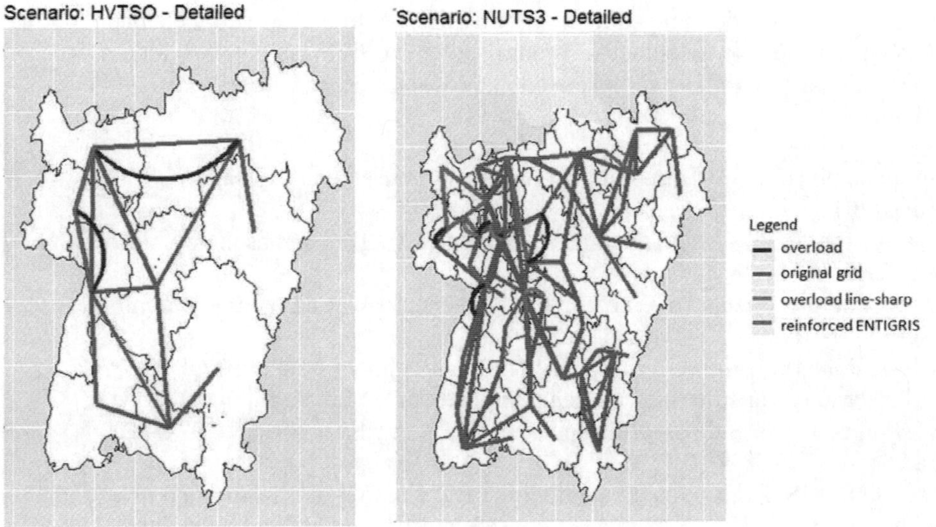

Figure 4: Grid expansions and detected overloads of DC load-flow model for the line sharp calculation of HVTSO (left and NUTS3_c right)

In the NUTS3_c calculation for both (aggregated and line sharp) cases, 10 upgrades are undertaken until 2050 in the ENTIGRIS model (marked in red). In comparison to the NUTS3_c case within the HVTSO case, 7 necessary grid upgrades were not detected because of the lower regional resolution in the optimization. Only in the aggregated model the years 2020 and 2030 result in a valid system. In all years of the line sharp case and in 2040 and 2050 in the aggregated case overloads occur that cannot be solved, because the line potentials are reached with the used model setup. Additionally three additional corridors are identified in the line sharp case in comparison to the aggregated case. In all calculated cases the expansions and the overloads occur in the North-West and in the North-East, in those corridors that were also expanded by the ENTIGRIS model.

5. Conclusion and Outlook

The paper shows that there are significant differences between a simple transport model and a DC load flow model. While the transport model optimizes the transportation by minimizing the costs, the DC load flow minimizes the electrical transport distance. The power flow on each represented line of both models deviate around 100 to 300 MW, with very rare occasions of more than 1 GW. The deviation of the line usage varies by around 10 %. The representation of transmission cost via normalized line length or reactance did not show an influence on the model results. Both approaches can be used.

Important for long term system planning is the grid expansion and the detection of critical lines. It can be concluded that an integral approach of optimizing the generation capacities, flexibilities and the grid infrastructure within one optimization problem is very beneficial because the social welfare can be increased using this approach. The analysis of the results shows that the ENTIGIS optimizer is able to identify most corridors that need reinforcement measures. The DC load flow approach can validate those results and specify which exact lines are critical. It was found that critical lines often occur when they are rather short in a meshed structure. For a long term system planning the approach seems reasonable.

Overall it can be concluded that the validation of the ENTIGRIS system optimizer with a DC load flow model are very valuable. Special attention need to be payed on the level of regional aggregation. A low regional resolution can give rough hints on necessary grid expansion measures whereas the accuracy increases by having a higher resolution. More specific expansion measures and cases of

IOP Conf. Series: Journal of Physics: Conf. Series **977** (2018) 012008 doi:10.1088/1742-6596/977/1/012008

overload can be detected and dealt with. While the DC load flow model should represent each single line in an investigated system, it is sufficient to use aggregated lines within the system optimizer. Further research needs to be done in defining for example time span as well as level of acceptable grid overload without upgrading the grid, as targeted redispatch measures for example are not considered within the DC load flow model. Additionally the accuracy of the model is also very dependent on the method that is used to cluster the grid. Methods of optimizing the overall system without the need of iterating need to be tested, to stay within acceptable computational times.

6. References

[1] BMUB. Klimaschutzplan 2050 - Klimaschutzpolitische Grundsätze und Ziele der Bundesregierung. 2015. http://www.bmub.bund.de/fileadmin/Daten_BMU/Download_PDF/Klimaschutz/klimaschutzplan_2050_bf.pdf. Accessed 27 Mar 2017.

[2] Bundesregierung. Energiekonzept für eine umweltschonende, zuverlässige und bezahlbare Energieversorgung; 2010.

[3] Krishnan V, Ho J, Hobbs BF, Liu AL, McCalley JD, Shahidehpour M, Zheng QP. Co-optimization of electricity transmission and generation resources for planning and policy analysis: Review of concepts and modeling approaches. Energy Syst. 2016;7:297–332. doi:10.1007/s12667-015-0158-4.

[4] Enzo E. Sauma, Shmuel S. Oren. Proactive planning and valuation of transmission investments in restructured electricity markets. Springer Science+Business Media, LLC 2006. 2006:358–87.

[5] Scholz Y. Renewable energy based electricity supply at low costs: Development of the REMix model and application for Europe; (Stuttgart: Universität Stuttgart)2012.

[6] Pfluger B. Assessment of least-cost pathways for decarbonising Europe's power supply. A model-based long-term scenario analysis accounting for the characteristics of renewable energies. (Karlsruhe: KIT Scientific Publishing); 2014.

[7] Hagspiel S. Cost-optimal power system extension under flow-based market coupling. Energy 2014; 66: 645-666.

[8] Kost CP. Renewable energy in North Africa: Modelling of future electricity scenarios and the impact on manufacturing and employment.(Stuttgart: Fraunhofer Verlag); 2015.

[9] Kost, Christoph; Junne, Tobias; Senkpiel, Charlotte; Hartmann, Niklas; Schlegl, Thomas; Zampara, Marilena; Capros, Pantelis. Renewable energy expansion and interaction in Europe: High resolution of RES potentials in energy system modeling; In: IEEE, editor. 12th International Conference on the European Energy Market (EEM); 19-22.05.2015; Lisbon, Portugal; 2015. doi: 10.1109/EEM.2015.7216677

[10] Senkpiel C, Shammugam S, Biener W, Hussein NS, Kost C, Kreifels N, Hauser W. Concept of evaluating chances and risks of grid autarky. In: IEEE, editor. 13th International Conference on the European Energy Market (EEM); 6-9 June 2016; Porto, Portugal; 2016. p. 1–5. doi:10.1109/EEM.2016.7521177.

[11] Glover JD, Sarma MS, OVERBYE TJ. Power system analysis and design. 5th ed. Stamford: Cengage Learning; 2012.

[12] Jonas Egerer, Clemens Gerbaulet, Casimir Lorenz. European Electricity Grid Infrastructure Expansion in a 2050 Context; 2013. (Berlin: DIW)

[13] Jonas Egerer. Open Source Electricity Model for Germany (ELMOD-DE): Data Documentation.(Berlin: DIW).

[14] eurostat. NUTS - Nomenclature of territorial units for statistics. 2017. http://ec.europa.eu/eurostat/web/nuts/overview. Accessed 18 Mar 2017.

ACKNOWLEDGEMENT

This paper is based upon work in the project AutGrid (No. 03EK3557A), within the initiative future-proof power grids supported by the Federal Ministry of Education and Research.

IOP Conf. Series: Journal of Physics: Conf. Series **977** (2018) 012009 doi:10.1088/1742-6596/977/1/012009

About the Need of Combining Power Market and Power Grid Model Results for Future Energy System Scenarios

Denis Mende, Diana Böttger, Lothar Löwer, Holger Becker, Alev Akbulut, Sebastian Stock

Fraunhofer Institute for Wind Energy and Energy System Technology IWES
Königstor 59, 34119 Kassel, Germany

denis.mende@iwes.fraunhofer.de

Abstract. The European power grid infrastructure faces various challenges due to the expansion of renewable energy sources (RES). To conduct investigations on interactions between power generation and the power grid, models for the power market as well as for the power grid are necessary. This paper describes the basic functionalities and working principles of both types of models as well as steps to couple power market results and the power grid model. The combination of these models is beneficial in terms of gaining realistic power flow scenarios in the grid model and of being able to pass back results of the power flow and restrictions to the market model. Focus is laid on the power grid model and possible application examples like algorithms in grid analysis, operation and dynamic equipment modelling.

1. Introduction

The expansion of renewable energy sources (RES) and the liberalisation of the energy markets in Germany and other European countries pose various challenges on the power grid infrastructure as power grid expansion is not keeping pace with increased transmission needs. The future energy system will even be based more on distributed RES generators instead of large conventional power plants. Additionally, sector coupling technologies (e. g. electric vehicles, heat pumps, power to gas) will have a big impact on the future power grid. For investigations on the interaction of electric power generation, grid and consumption, models for the power market as well as the power grid are necessary. At Fraunhofer Institute for Wind Energy and Energy System Technology (IWES) models for the power market and the transmission grid have been developed separately from each other. Now they are coupled by exchanging data and simulation results to better and fully address these challenges [1].

In literature, there are several approaches to model power generation and power grid as well as their interaction. Most market models assume that there are no technical restrictions or limits in transmission capacity within the grid. Exchange capacities are only taken into account between single countries or market regions. In contrast, detailed power grid models spend special focus on grid restrictions and bottlenecks addressing research questions like grid planning, price zone tailoring, exchange capacities or optimal power plant operation in terms of the status of the power grid itself and grid operator requirements. In order to take future developments of the power plant fleet and the respective degree of capacity utilisation of single power plants into account, it is necessary to analyse the interaction between power market and power grid. For example, an integrated power market and power grid model is used to calculate an optimal market area tailoring in central Europe [2]. In contrast, a power grid model only covering Germany is used to investigate the feasibility of regional energy autarky [3]. Additionally, power grid modelling can pursue different approaches by using the more complex alternating current (AC) load flow as most of the above mentioned models do, or a simplified, so called direct current (DC) load flow approach, as e. g. presented in [4]. A broad overview of respective literature, models and their implementation is given in [1].

This paper is organised as follows: In the second and third chapter the IWES power market model and the power grid model will be introduced before the model combination is described in chapter 4. Chapter 5 gives an outlook especially on future applications of the power grid model and chapter 6 summarises the paper and gives an outlook on further works.

2. Power Market Model

The model family "SCOPE" developed at IWES encompasses fundamental energy system models for generation expansion planning as well as for power plant dispatch. The basic model was developed and is explained in more detail in [5]-[6]. The capacity expansion model part is used to determine newly installed capacities of power plants, RES, storages, transfer capacities as well as heat technologies as linear optimisation problem (see Figure 1). These results are passed to the detailed dispatch model part considering the demand for balancing power as well as the heat demand and the transportation sector requirements in more detail (mixed-integer optimization). The power market model considers any desired period (usually one year) in hourly time steps. Regarding the electricity, heat and transportation sector as well as the coupling of them, SCOPE covers Europe. The heat sector encompasses the demand for space heating, domestic hot water and process heat. The transport sector covers an hourly representation of the (within certain limits) flexible power demand of electric vehicles (battery electric, plug-in hybrid, range extender and eHighway trucks). Thereby, power plant fleet specific input data with a high spatial resolution is used, whereas restrictions in transmission are just incorporated as transfer capacities between countries or market areas in this step. As a result, the power market model calculates unit-wise power plant schedules. For more detailed descriptions see [5]-[8].

An example of a future power generation scenario for the year 2025 was designed to consider sector coupling as follows ([9], compare also [1]): The installed capacities of RES and thermal power plants are mainly based on scenarios used by transmission system operators for their grid expansion planning [10]-[11]. Therefore, they describe realistic assumptions concerning the future development of the energy system. For the investigations, the data of the scenario "B" of [10] was linearly interpolated for the year 2025. Fossil-thermal power plants are assumed to be decommissioned according to their technical life time while combined heat and power plants (CHP) are replaced by new gas-fired CHP plants. Power generation profiles of wind and solar power plants are derived by own models. The power consumption of Europe was taken from the scenario "EU 2020" of [12] also describing a realistic forecast of the European transmission system operators. Figure 2 shows the power generation of renewable and conventional energy sources for two illustrative weeks in August 2025. The situation is characterised by high peaks of photovoltaic (PV) generation (upper part of the figure). Additionally, there is high wind power generation in the second week. While the power generation of RES is below the power demand in the first week, RES generation exceeds the demand in the following week. Dur-

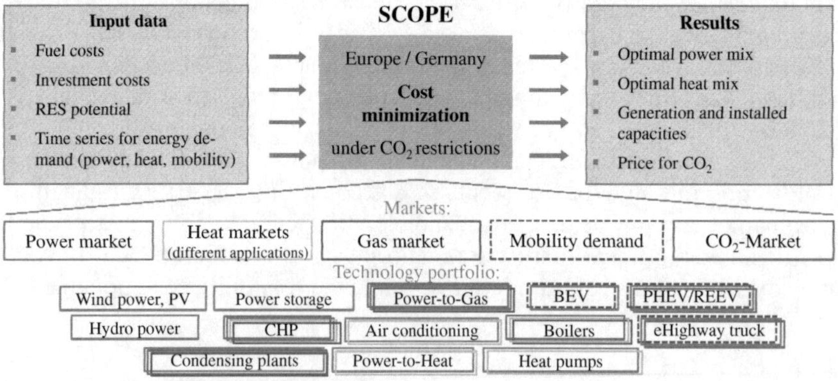

Figure 1. Overview of the model SCOPE for investment and dispatch optimisation across energy sectors (colours of the technology boxes refer to the associated markets).

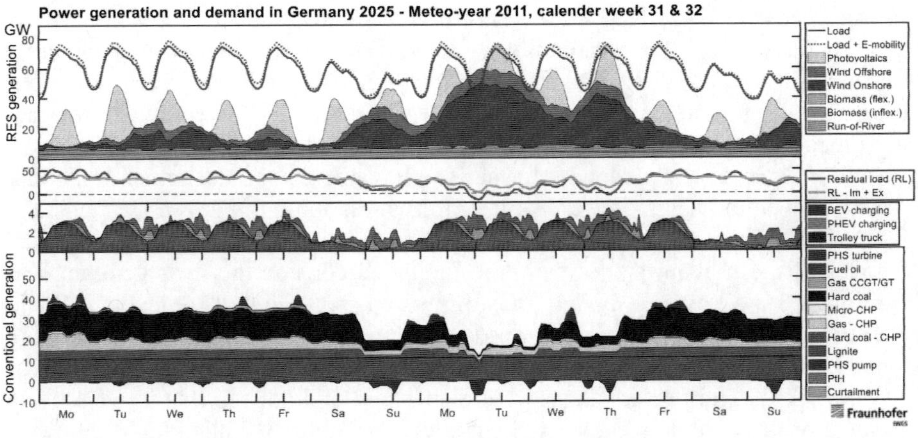

Figure 2. Generation of RES in comparison to load (upper figure). Residual load and power consumption of the transportation sector (middle). Dispatch of power plants (lower figure) for two weeks in August 2025 [1].

ing this time the power generation of thermal power plants is decreased considerably (lower part of the figure). During the highest RES generation only lignite power plants are still running because of the provision of system services. Additionally, pump-storage hydro power plants are dispatched to store the excess power on a daily scale. The power export of Germany is quite high during times with low residual load. In the middle part of the figure, the power consumption of the transportation sector is depicted. While the demand of eHighway (trolley) trucks is relatively inflexible, the charging profile of passenger cars can follow the RES-generation using residual load dependent charging schemes. From the model results it can be seen that lignite substitutes the base load of former nuclear power plants while other fossil-thermal power plants need to produce power more flexibly. Accordingly, the power flows in the power grid will change depending on the RES in-feed situation.

3. Power Grid Model

The current version of the power grid model, as introduced in [1], which was originally gathered at the Institute of Electric Power Systems of Leibniz University Hanover, uses publicly available data of the European power system as given in respective ENTSO-E grid maps on the extra high voltage levels 220 and 400 kV [13]-[14]. At IWES, this power grid data, as developed in [13]-[14], is available for the continental part of the ENTSO-E region and with higher level of detail for the German part of the grid (see Figure 3). Since accurate power grid modelling especially is an issue of data availability and quality, the major problem in the past (compare [15]), validation and enhancements of the current data is important. Based on an increasing availability of such data due to open source mapping services such as [16]-[17] and initiatives to extract and interpret this data [15], the access to real electrical grid data in general has improved.

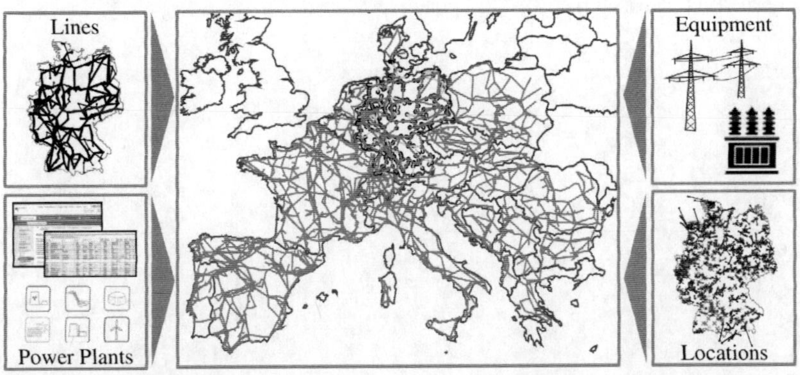

Figure 3. Components, available data and currently realised transmission system model (see also [1]).

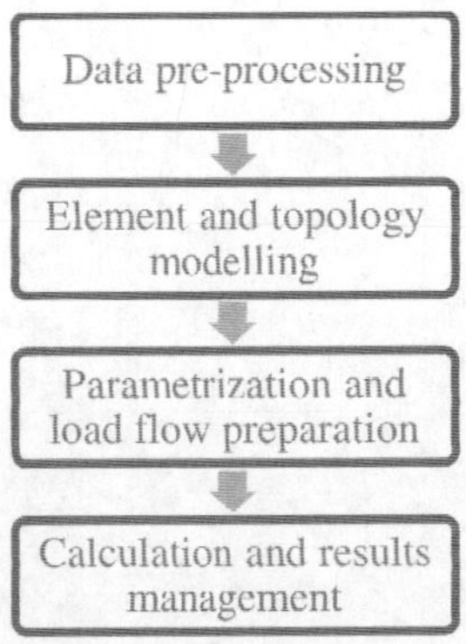

Figure 4. Automated steps in modelling and load flow.

The commercial software package DIgSILENT PowerFactory was chosen for the modelling in order to utilise the graphical representation of the grid topology as well as its implemented analysis functionalities. In order to be able to generate different grid scenarios (e. g. diverse future years or various scenarios for the same year), an automated approach to generate the model in PowerFactory was developed. Therefore, a data pre-processing using MATLAB was realised, in which grid scenarios, equipment data and generation units as well as load information are gathered and the actual modelling is prepared. The grid itself is modelled using the respective information about the grid's elements and topology as given in [13]-[14]. The transmission system consists of lines and nodes with typical equipment types. The nodes of the transmission system are modelled considering their geographical position (see "Lines" in Figure 3). Within the model, transformers connect the voltage levels 220 and 400 kV of the transmission grid. All nodes were equipped with loads that represent the underlying network structures. Conventional power plant data was implemented using the list of power plants of the Federal Network Agency [18]. The power plants were distributed and connected to nodes using information about their locations as shown in the part "Locations" of Figure 3. Following, the grid is parameterised and assumptions are set for the load flow (or other) calculations. These steps as well as the actual load flow calculation are automated using the DIgSILENT Programming Language (DPL). Finally, the calculation results are post-processed using MATLAB to generate respective figures and diagrams. The whole process is illustrated in Figure 4.

A typical application example using the power grid model is the redispatch analysis, which was performed and presented in [19]-[20]. A quantitative/technically optimal ("tech") as well as cost-effective ("eco") solution of an optimisation problem for the redispatch of power plants is presented. Furthermore, process enhancements allowing a reasonable trade-off between the solutions of quantity P_{tech} and cost effectiveness C_{eco} are developed. This was realised by normalisation ("norm") of the results of the previous optimizations (P_{tech} and C_{eco}) in order to find the minimal deviation from the ideal point $\langle C_{eco}|P_{tech}\rangle$ of redispatched power and costs (for further explanation on the normalization process, see [19]). For the analysis, power plant dispatch as well as load was simply evenly distributed among the power plants. Slight changes in the load distribution and dispatch lead to a congestion presenting a typical situation in the German grid [21]. Costs were assumed for dispatch changes of the power plants as variable and as fixed costs.

Relating the different optimisation targets "tech", "eco" and "norm", the results are shown in Figure 5 (see also [20]). Therein, "before" and "after" show the line loading according to colours from green (0 % loading) to red (100 % loaded) while the figures in the middle show the locations and volumes of the redispatched power according to the circle sizes given in orange (decrease of generation) and blue (increase in generation). All three variants lead to a grid without congestion (see "after"). While the "eco" optimisation leads to the lowest costs, the amount of redispatched power is higher compared to the "tech" case which on the other hand leads to much higher costs as the "eco" approach. In comparison, "norm" leads to only slightly higher costs than the "eco" case and only a marginal increased amount of redispatch power compared to the "tech" approach.

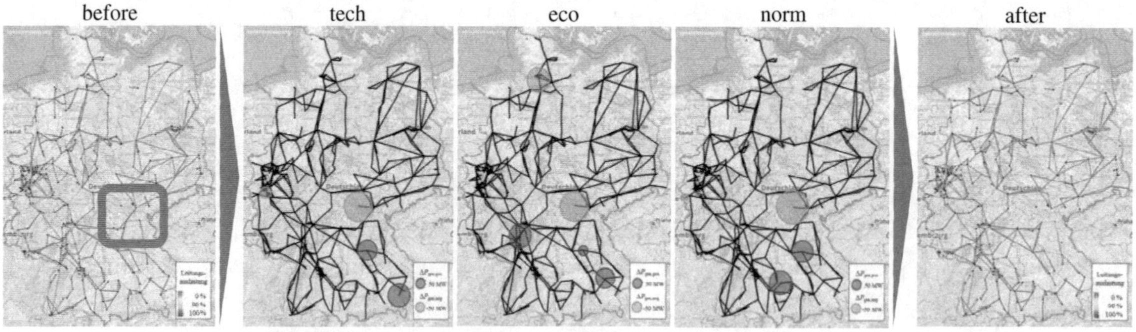

Figure 5. Results of the application example congestion management using redispatch of power plants with different optimisation targets (for further information see [20]).

IOP Conf. Series: Journal of Physics: Conf. Series **977** (2018) 012009 doi:10.1088/1742-6596/977/1/012009

4. Combination of Models

For research questions it is essential to simulate future scenarios of the power market and grid. In order to be able to simulate realistic scenarios and to analyse the behaviour of market mechanisms and technical challenges, a combination of the models is necessary. The current implementations follow a sequential approach, which represents today's procedure in reality. This allows the analysis of realistic power plant and renewable feed-in and therefore occurring technical challenges in the grid. Additionally, resulting constraints which are detected via the grid calculation(s) can be used as feedback for the market model, especially in terms of price zones, transfer capacities or further technical restrictions.

Up to now, the market result of the power plant dispatch is transferred to the power grid model providing set points for each single power plant. In order to realise realistic power flows within the grid model, also loads and renewable energy generation has to be distributed among the nodes of the transmission grid which currently is in progress. The market model generates hourly values for any desired time range (e. g. one year). Since market model and power grid model interact by exchanging simulation results, but generally are working on their own, there is just a soft-link between them.

One first application showing the combination of the developed power market and grid model combination is a contingency analysis. For this purpose, market model results for power plant in-feeds for the historical year 2011 were implemented in the power grid model (see also [1]). The market results represent a precise power plant dispatch on an hourly basis for the whole year. For the analysis, the residual load – as a further result of the market model – was simply evenly distributed to the network nodes neglecting regional differences in RES in-feed and power demand. Power exchange to neighbouring market areas or countries were considered in the market model, but not explicitly modelled in the grid. For the power flow calculation all generators were set to voltage control, while typical active and reactive power limits of the elements were deployed. The set point for the voltage control was set to 1.03 per unit (p.u.).

The results of the contingency analysis are presented in Figure 6, in which part a) shows the power generation of dispatchable plants and the residual load. The map in b-1) shows the average loading of the lines represented by width and colouring showing the percentage of overloaded time during the simulated year. Depicted are the average loading of the lines (line width – green) and the congested time (loading above rated/threshold value) ranging from no overloading (0 % – yellow) to congested throughout the year (100 % red). Part b-2) gives the average node voltages as percentage from nominal voltage (100 %). Typically a voltage range of +/-10 % around nominal voltage is acceptable and also occurring in this example. Therefore the scale is given from 90 to 110 % (of nominal voltage). Since the results represent a unit-wise dispatch for the conventional power plants while the residual load (load minus renewable power generation) is just evenly distributed among the nodes, the presented results do not show typical overloadings as observed and reported in the recent years (see e. g. [21]). Nevertheless, underlying methods and algorithms are developed and subsequently can be used with more realistic assumptions. Also, a realistic distribution of load as well as RES is in development.

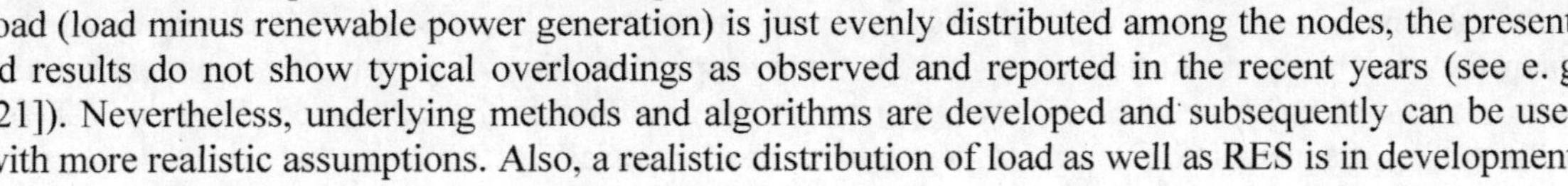

Figure 6. a) Power plant dispatch and residual load as result of the market model. b-1) Average line loading (line width) and congestion as percentage of time of the year (coloured from yellow to red). b-2) Average node voltages as percentage of nominal voltage for a years' power plant dispatch.

5. Further applications

Within the following, further typical applications in the categories grid analysis & operation and dynamic modelling are presented. Most of these applications are developed in small test systems or for grids with just a few nodes showing the respective behaviour. In order to generate statements concerning the future behaviour in the German and European power system, these algorithms and control schemes have to be implemented in or at least scaled to the introduced models of the power system. Since system responsibility is an issue of transmission system operators (TSO), questions of grid analysis and operation, dynamic behaviour of active grid components and also system restoration have to be modelled and investigated on transmission system level.

Questions of market mechanisms on the one hand and technical boundaries and dependencies on the other hand need respective modelling. Being able to investigate questions on either and especially on both sides calls for models that are able to represent the whole picture. Analysis of power grid challenges can only be addressed adequately, if the power flow scenario – which directly is defined through generation and load distribution – is realistic or describes a certain situation. Therefore investigations on power grid side require the results of the power market. On the other hand, the liberalized power market is constrained by technical boundaries such as transfer capacities, which only can be included correctly, if an adequate power grid model allows to determine these demands correctly. In the following, several application examples are given, which take market as well as technical modelling into account.

5.1. Power system analysis & operation

Active as well as reactive power management for the transmission grid, especially in the context of changing supply structures and decentralised generation in the transmission connected distribution grids, is a field of research with large impact on system security and efficiency. Addressing this issue, optimisation algorithms for reactive power management in transmission and distribution grids are possible solutions. In [22] a control concept is proposed using reactive power capabilities of wind farms and tap-changer positions of transformers to improve distribution grid operation. Control signals, namely tap-changer positions and reactive power set-points, are smoothed over the forecast horizon using the implemented algorithms. Further possible optimisation objectives are power loss reduction, voltage profile smoothing and complying with reactive power exchange limits with the transmission grid. Being able to number the improvements needs correct scenarios (derived out of market models) as well as correct technical representations (on grid model side).

Driven by the liberalisation of the electricity market as well as cross border energy trade and the large-scale integration of renewable energy sources, electrical grids are subject to power flows, for which they are not planned and yet prepared. Since implementation of grid enforcement plans take long time, coordinated methods and management of congestion must be developed. Especially in future scenarios the installation of new wind farms on- and especially offshore leads to further challenges. Furthermore, the impact of voltage source converters in high voltage direct current (HVDC) transmission systems on economically optimal redispatch scenarios is a field of research for which powerful models of realistic scenarios derived from market structures and the modelling of respective technical equipment on power grid side are necessary.

Additional works incorporating HVDC systems consider offshore networks, their long term planning and layout taking optimal market as well as power grid integration into account. Balancing and congestion management on- and offshore requires respective models of the considered areas and applications.

5.2. Power system dynamics

Frequency control in interconnected electrical power systems is nowadays realised using the networks' conventional thermal power plants. Up to now a contribution of renewable sources to primary frequency control is not realised in interconnected power systems. Possibilities to support the frequency of power systems using wind turbines with enhanced active power functionalities such as primary frequency control and synthetic inertia are presented e. g. in [23]. In future works, these investigations could be implemented in models of real power systems using generation patterns derived from the market model incorporating weather data for RES.

In todays and future power systems, HVDC is to be used in electrical power systems to increase transmission capacity, contributing to enhance the system stability, security and controllability. Thus, HVDC is applied to transfer bulky power over long distances. One of the application areas in this sense is the consideration of HVDC systems and their flexible controllability in the event of power system restoration. Operating HVDC systems in these scenarios, developing methods and algorithms to generate set points and control structures needs input of realistic boundary conditions in power systems and therefore respective models.

Today's dynamic models for stability simulations mainly focus on the electrical part of the generator of conventional power plants. Compared to this, active power dynamics during medium time events (in the range of a couple of seconds to a few minutes) is mainly influenced by the steam turbine. Usually the upstream steam process and the pressure control system are not modelled. In many cases it is simply assumed that steam power is available. To conduct dynamic simulations in applications with small or islanded systems on a time scale not only including a few seconds after a disturbance, consideration of energy storages within the system is important. These storages have to be considered, since they provide the power to support the frequency after load connections. In combination with grid models, the investigation of the dynamic behaviour also in ranges of several minutes becomes possible.

6. Summary & further work

After a brief introduction of models for power market and power grid simulation developed at IWES, the combination of these two models and occurring tasks and challenges are presented. Typical applications of the single models, as well as a first analysis of the combination of market and power system model are shown. Using the models it is possible to simulate the (future) energy system starting from a power markets' perspective up to typical questions in grid operation and planning. Especially the integrated approach of power, heat and transportation allows analysing future energy scenarios.

Further possible applications are presented in the fields of grid operation and dynamic modelling. In the field of grid operation the modelling of real power systems is important to develop, test and validate algorithms and optimisation schemes to be used in grid operation and analysis. Furthermore, possible applications of dynamic models of network equipment have to be combined with grid structures in order to analyse the dynamic behaviour in small time ranges (seconds) as well as over times going up to several minutes and hours.

Further applications and research especially cover system studies on the implementation of future energy scenarios, the impact of new technologies as well as technical and algorithmic aspects on transmission grid level. Furthermore, especially system restoration is not an issue of daily work and cannot be trained in real systems, powerful simulation tools and respective models are important to ensure and validate grid restoration plans and to enable training for operators.

Acknowledgments

This research work was supported by the German Federal Ministry for Economic Affairs and Energy and the Projektträger Jülich GmbH (PTJ) within the framework of the project SystemKontext (FKZ: 03240010).

References

[1] Mende D, Böttger D, Ganal I, Löwer L, Harms Y, Bofinger S 2017 Combined power market and power grid modeling – First results of the project SystemKontext *14th International Conference on the European Energy Market* Dresden (to be published).

[2] Deilen C, Felling T, Felten B, Niesen L, Osinski P, Weber C 2016 KoNeMaSim – Kopplung von Netz- und Marktsimulation für die Netzplanung *Conference Zukunftsfähige Stromnetze* Berlin.

[3] Senkpiel C, Schammugam S, Biener W, Hauser W, Kost C, Kreifels N, Linke M, Krüger F 2016 Autgrid:

Chancen und Risiken der Netzautarkie in Deutschland *Conference Zukunftsfähige Stromnetze* Berlin.

[4] Leuthold F, Weigt H, v. Hirschhausen C 2012 A Large-Scale Spatial Optimization Model of the European Electricity Market *Netw Spat Econ* **12** pp 75-107. DOI: 10.1007/s11067-010-9148-1.

[5] v. Oehsen A 2014 Entwicklung und Anwendung einer Kraftwerks- und Speichereinsatzoptimierung für die Untersuchung von Energieversorgungsszenarien mit hohem Anteil erneuerbarer Energien in Deutschland *Dissertation* University of Kassel.

[6] Jentsch M 2014 Potenziale von Power-to-Gas Energiespeichern. Modellbasierte Analyse des markt- und netzseitigen Einsatzes im zukünftigen Stromversorgungssystem *Dissertation* University of Kassel.

[7] Fraunhofer IWES 2016 SCOPE: Sektorenübergreifende Einsatz- und Ausbauoptimierung für Analysen des zukünftigen Energieversorgungssystems. *Retrieved from* http://www.energiesystem technik.iwes.fraunhofer.de/content/dam/iwes-neu/energiesystemtechnik/de/Dokumente/Broschueren /2015_F_SCOPE_web.pdf (November 24, 2016).

[8] Trost T 2016 Erneuerbare Mobilität im motorisierten Individualverkehr. Modellgestützte Szenarioanalyse der Marktdiffusion alternativer Fahrzeugantriebe und deren Auswirkungen auf das Energieversorgungssystem *Dissertation* University of Leipzig.

[9] Fraunhofer IWES 2015 Analyse und Darstellung der Klimawirksamkeit der Elektromobilität in zukünftigen Stromversorgungsszenarien. Szenariensimulation, Analyse und Darstellung. Ableitung von optimalen trukturellen Entwicklungspfaden für den Verkehrs-und Wärmesektor. *Retrieved from* http://projekt-envelmo.iwes.fraunhofer.de/includes/reports/Endbericht_Stromversorgungsszenarien_ EMobilitaet_IWES.pdf (March 24, 2017).

[10] ENTSO-E 2014 Scenario Outlook & Adequacy Forecast 2013-2030 - Dataset. *Retrieved from* https://www.entsoe.eu/outlooks/maf/Pages/default.aspx (October 1, 2014).

[11] 50Hertz Transmission GmbH, Amprion GmbH, TenneT TSO GmbH, TransnetBW GmbH 2014 Szenariorahmen zum NEP/ O-NEP 2015 – Entwurf. *Retrieved from* http://www.netzentwicklungs plan.de/begleitdokumente/2015 (October 1, 2014).

[12] ENTSO-E 2013 Scenario Outlook and Adequacy Forecast 2012 – 2030. *Retrieved from* https://www.en tsoe.eu/fileadmin/user_upload/_library/SDC/SOAF/120705_SOAF2012_final.pdf (October 1, 2014).

[13] Rathke C 2013 Entwicklung eines Modells für die integrierte Simulation der europäischen Übertragungsnetze und Strommärkte *Dissertation* Leibniz Universität Hannover.

[14] Rendel T 2015 Erweiterung und Plausibilisierung eines Modells für die integrierte Simulation des europäischen Verbundnetzes und Strommarktes *Dissertation* Leibniz Universität Hannover.

[15] Medjroubi W, Müller U P, Scharf M, Matke C, Kleinhans D 2017 Open Data in Power Grid Modelling: New Approaches Towards Transparent Grid Models *Energy Reports* **3** pp 14-21. DOI: 10.1016/j.egyr.2016.12.001.

[16] http://www.openstreetmap.org (February 8, 2017).

[17] http://www.flosm.de/html/Stromnetz.html (February 8, 2017).

[18] Bundesnetzagentur 2016 List of Power Plants. *Retrieved from* http://www.bundesnetzagentur.de/ cln_1422/EN/Areas/Energy/Companies/SecurityOfSupply/GeneratingCapacity/PowerPlantList/ PubliPowerPlantList_node.html (November 24, 2016).

[19] Mende D, Stock S, Hennig T, Löwer L, Hofmann L 2016 Multiobjective Optimization in Congestion Management Considering Technical and Economic Aspects *IEEE PES Asia-Pacific Power and Energy Engineering Conference* Xi'an.

[20] Mende D, Stock S, Hennig T, Löwer L, Hofmann L 2016 Effizientes Engpassmanagement im Spannungsfeld von technischer und wirtschaftlicher Optimierung *VDE-Kongress 2016* Mannheim.

[21] Bundesnetzagentur 2016 Quartalsbericht zu Netz- und Systemsicherheitsmaßnahmen – Ersters Quartal 2016. *Retrieved from* https://www.bundesnetzagentur.de/EN/Areas/Energy/Companies/ SecurityOfSupply/NetworkSecurity/Network_security_node.html (February 13, 2017).

[22] Stock S, Venzke A, Hennig T, Hofmann L 2016 Model Predictive Control for Reactive Power Management in Transmission Connected Distribution Grids *IEEE PES Asia-Pacific Power and Energy Engineering Conference 2016* Xian pp 419-423. DOI: 10.1109/APPEEC.2016.7779538.

[23] Mende D, Hennig T, Akbulut A, Becker H, Hofmann L 2016 Dynamic Frequency Support with DFIG Wind Turbines – A System Study *IEEE Electrical Power and Engineering Conference* Ottawa. DOI: 10.1109/EPEC.2016.7771694.

Institute of Physics
Dirac House, Temple Back
Bristol BS1 6BE UK

ISSN: 1742-6588
ISBN 978-1-5108-5954-8

CURRAN ASSOCIATES INC.
proceedings.com